SECOND EDITION

CHANGING CONTOURS OF WORK

Other Titles for Your Consideration From the Sociology for a New Century Series

Sociology and Human Rights: A Bill of Rights for the Twenty-First Century, by Judith Blau and Mark Frezzo

Beyond a Border: The Causes and Consequences of Contemporary Immigration, by Peter Kivisto and Thomas Faist

Economy/Society, Second Edition, by Bruce Carruthers and Sarah Babb

Cities in a World Economy, Fourth Edition, by Saskia Sassen

Gods in the Global Village: The World's Religions in Sociological Perspective, Third Edition, by Lester R. Kurtz

Cultures and Societies in a Changing World, Fourth Edition, by Wendy Griswold

How Societies Change, Second Edition, by Daniel Chirot

The Sociology of Childhood, Third Edition, by William A. Corsaro

Constructing Social Research: The Unity and Diversity of Method, Third Edition, by Charles C. Ragin and Lisa M. Amoroso

Women, Politics, and Power: A Global Perspective, by Pamela Paxton and Melanie Hughes

SOCIOLOGY FOR A NEW CENTURY

SECOND EDITION

CHANGING CONTOURS OF WORK

Jobs and Opportunities in the New Economy

STEPHEN SWEET

Ithaca College

PETER MEIKSINS

Cleveland State University, Cleveland, Ohio

Los Angeles | London | New Delhi
Singapore | Washington DC

Los Angeles | London | New Delhi
Singapore | Washington DC

FOR INFORMATION:

SAGE Publications, Inc.
2455 Teller Road
Thousand Oaks, California 91320
E-mail: order@sagepub.com

SAGE Publications Ltd.
1 Oliver's Yard
55 City Road
London, EC1Y 1SP
United Kingdom

SAGE Publications India Pvt. Ltd.
B 1/I 1 Mohan Cooperative Industrial Area
Mathura Road, New Delhi 110 044
India

SAGE Publications Asia-Pacific Pte. Ltd.
3 Church Street
#10-04 Samsung Hub
Singapore 049483

Acquisitions Editor: Dave Repetto
Editorial Assistant: Lydia Balian
Production Editor: Eric Garner
Copy Editor: Trey Thoelke
Typesetter: Hurix Systems Pvt. Ltd.
Proofreader: Laura Webb
Indexer: Rick Hurd
Cover Designer: Candice Harman
Marketing Manager: Jordan Bell
Permissions Editor: Karen Ehrmann

Printed in the United States of America.

Library of Congress Cataloging-in-Publication Data

Sweet, Stephen A.

Changing contours of work : jobs and opportunities in the new economy / Stephen Sweet, Peter Meiksins.—2nd ed.

p. cm.—(Sociology for a new century)

Includes bibliographical references and index.

ISBN 978-1-4129-9086-8 (pbk.)

1. Technological innovations—Economic aspects—United States. 2. Labor market—United States. 3. High technology industries—United States. 4. Hours of labor—United States. 5. Industrial relations—United States. 6. United States—Economic conditions—21st century. 7. Globalization. I. Meiksins, Peter, 1953- II. Title.

HC110.T4S93 2013

331.0973—dc23

2012005030

This book is printed on acid-free paper.

MIX
Paper from responsible sources
FSC
www.fsc.org
FSC® C014174

12 13 14 15 16 10 9 8 7 6 5 4 3 2 1

Contents

To our hard-working students

List of Exhibits

About the Authors

Stephen Sweet is an Associate Professor of Sociology at Ithaca College. His studies of work, and its impact on and off the job, have appeared in a variety of publications, including *Work and Occupations; Sex Roles; Research in the Sociology of Work; Family Relations; New Directions in Life Course Research; Journal of Vocational Behavior; Journal of Marriage and the Family; Innovative Higher Education; Teaching Sociology; Generations;* and *Community, Work, & Family.* His books, *College and Society: An Introduction to the Sociological Imagination* and *Data Analysis With SPSS: A First Course in Applied Statistics* have been extensively adopted in sociology courses. He has edited and coedited publications including the *Work and Family Handbook: Interdisciplinary Perspectives, Methods, and Approaches; The Work and Family Encyclopedia;* and special issues of the journals *Teaching Sociology* and *Community, Work & Family.* His current research focuses on the factors that lead people to embrace work as part of their identity and the ways that organizations integrate flexible work practices. In his job at home, he provides support to (and receives support from) his wife Jai (a college administrator) and children Arjun and Nisha.

Peter Meiksins is Professor of Sociology and Director of the Honors Program at Cleveland State University. He has published widely on the sociology of work, particularly the sociology of technical work and of the professions, in journals such as *Work and Occupations; Sociological Quarterly; Work, Employment & Society; Theory and Society; Technology and Culture;* and *Labor Studies Journal.* He has coedited several books on work and labor and is the coauthor of two books: *Engineering Labour: Technical Workers in Comparative Perspective* and *Putting Work in its Place: A Quiet Revolution.* His current research focuses on job dislocation in STEM fields and on gender and engineering. He has a daughter in university and he does all the cooking at home (he likes to cook; his wife doesn't); his wife barbecues and fixes the car.

Preface

This book is an effort to make sense of work opportunity—as it was in the 20th century and as it is today—and how it influences lives on and off the job. When we began writing the first edition of this book, we thought that this would be a straightforward endeavor. First, we intended to discuss the "old economy" and the types of opportunities present when most of the labor force was employed in jobs critical to mass production industrial work. Then we were going to write about the emerging "new economy" and the ways new technologies, new organizations, new jobs, a new workforce, and globalization are transforming work. Our unique contribution would be to show the ways that current policies and practices, designed to correspond with needs in the old economy, fail to address the present-day concerns of working families.

When we wrote the first edition, we spent well over a year blocking out chapters, going back into the research literature, writing chapter drafts, restructuring our arguments, and rewriting. But we faced a recurring problem, namely, that our observations about the old economy kept intruding into what we wanted to say about the new economy, and vice-versa. Our work in that first year would have been far easier if we had recognized then what is now a central theme of this book: *the old economy has not been replaced by a new economy; the old economy is operating within the new economy.*

Once we recognized this, we realized that our thesis would have to be modified, as would the structure of our project. The story of the old and new economies is one of *common social forces* that shape the development of work opportunity. Many features of the old economy, although sometimes in new forms, are central to the dynamics of the new. Our conclusion is that concerns facing workers today result from *structural lags* that have forestalled the implementation of effective responses to changes in the ways work is performed and from *enduring failures* to address the problems of inequality that developed in the old economy.

In the years that followed the publication of the first edition, the global economy tanked and the housing bubble burst. Job insecurity expanded, homes were lost, and working families experienced compounded strains. But not everything that happened was bad. There were some important expansions in workers' rights, such as increased opportunity to file discrimination suits and expanded opportunities for working women to breastfeed their children. But perhaps the biggest story, in respect to work and opportunity, concerns reckless decisions made to serve the interests of those at the very top of the opportunity ladder, and the consequences those decisions had on almost everyone else. While the observations that we presented in the previous edition remain largely the same, new and updated statistics presented in this edition show that opportunity divides have continued to expand, rather than contract. It is abundantly clear that the new economy is not working for everyone.

In the chapters that follow, our goals are to identify the contours of work and how they have changed over time, considering both short-term changes that may have occurred over the course of the preceding decades, as well as the longer-term development of modern ways of organizing work. Our analysis relies primarily on the research of sociologists, but also on that of labor historians and economists. Our goal is not to offer comprehensive histories of work, or to detail the experiences of all groups in the workforce, but to document the processes that shape work opportunity and how opportunities have been divided in the United States along class, gender, and racial lines. To do this, we adopted a comparative perspective, placing our analysis of opportunity and policy in the United States alongside the somewhat different realities of work in Western Europe and elsewhere. We also compare the experience of workers laboring today with those laboring in the mid-20th century and earlier, and we explore the American workplace in the larger context of an integrated global economy and emerging global networks of trade.

Chapter 1, "Mapping the Contours of Work," offers an introduction to the sociology of work and the unique contributions sociological analysis brings to the understanding of the changing economy. Our concern in this chapter is not so much to detail the nature of work in the new economy, or how changes in work have happened, but rather to indicate what needs to be examined if one is to understand work, society, and social change today. To do this, we outline observations sociologists have made about the ways culture, social structure, and agency shape the opportunity to work and the careers of workers. We introduce this chapter by describing the challenges faced by five different workers laboring in the new economy. These individuals illuminate the *diversity* of workers' experiences, and how the transition to a new economy is affecting career prospects and introducing distinct strains into family lives.

Chapter 2, "New Products, New Ways of Working, and the New Economy," considers the changing patterns of what is produced and how production occurs. In this chapter we consider the implications of concerns such as deindustrialization, the rise of service sector employment, as well as changing organizational designs and technologies. The primary question we consider here is the extent to which the new economy differs from the old economy in respect to what is created and the labor processes and practices involved in production.

Chapter 3, "Economic Inequality, Social Mobility, and the New Economy," examines the economic returns received from work and how work opportunity gives shape to the class structure of society. The analysis reveals sobering signs that economic transformations are contributing to a divided economy, one that sustains a two-tiered division between good jobs and bad jobs, and one that is funneling substantial shares of the returns of work to a privileged elite. We also consider how the movement of "good jobs" from the United States affects the life chances of workers in emerging economies, as well as more peripheral areas of the global economy.

Chapter 4, "Whose Jobs Are Secure?" and Chapter 5, "A Fair Day's Work? The Intensity and Scheduling of Jobs in the New Economy," consider how security and time commitments to work have changed. We first show the ways changing work designs in the new economy are contributing to widening job insecurity. Our interest here is not just to detail the extent of risk present today, but also to show how social policies implemented in the old economy set workers up to bear the burden of risk, often at the expense of their families and careers. Chapter 5 extends this history of the present by examining trends in the time spent working and the intensity of work. Here, we discuss the question of why American workers are working more than they did in the past, more than workers in almost every other society, and in many instances more than they want to. We also consider the implications of work in a 24/7 economy and the impact nonstandard schedules have on family lives.

Chapter 6, "Gender Chasms in the New Economy," examines the issue of gender inequalities at work. We revisit the fundamental question of what constitutes work, and why women's contributions to society are commonly defined as something other than "real work" or not worthy of compensation commensurate to that received by men. We also consider the extent to which gender inequalities are disappearing in the new economy and detail why many inequalities persist. We conclude this chapter by examining the approach to handling care work in the United States, how it departs from the approaches used in Western Europe, and its impact on both the quality of care and women's life chances.

Chapter 7, "Race, Ethnicity, and Work: Legacies of the Past and Problems in the Present," examines the proposition that race might be of declining significance in the new economy. We show that racial inequalities persist but that there are important differences in the ways various minority groups have responded to, and are being treated in, the new economy. We also detail the dominant reasons why racial and ethnic inequalities exist today. Because race continues to be a major policy concern, we consider two of the most pressing debates: the controversies about affirmative action programs and about the impact of immigration on opportunity structures.

Chapter 8, "Reshaping the Contours of the New Economy," outlines what needs to change if work is to become a positive experience for all and how opportunities might be distributed more equitably in the new economy. Basing our recommendations on what has been done in other developed societies, we try to offer realistic goals that, if fulfilled, would create more humane workplaces. But we also acknowledge that the dehumanizing, unjust aspects of work in the new economy are unlikely to change by themselves and that positive steps must be taken to promote improvements. A variety of agents—including individuals, activist groups, unions, corporations, and government entities—will all need to play a role in reshaping work. In the end, we suggest that government intervention will be the key to bringing the expectations of employers in line with what should be expected of workers. Its level of engagement will hinge on the ability of individuals, activist groups, and unions to exert sufficient pressures.

Our hope is that this book will help readers to understand the origins of current problems confronting working people in the new economy. Beyond this, we hope this book will contribute to a much-needed dialogue about the strategies for liberating workers from poverty, from drudgery, from discrimination, from stress, and from exploitation.

Acknowledgments

This book is the result of the contributions of numerous people, from those who cut the trees, milled them into paper, drove the paper to our offices, designed our computers, filled our libraries with books (or at least electronic links to books!), fed us and our children, heated our offices, and to those who printed and delivered it to your hands. Our intellectual efforts stand on the shoulders of the giants in the field, individuals who introduced the ideas we tried to advance and to whom our thoughts are indebted. We also relied on the efforts of the numerous researchers and officials who collected and organized the data we use to outline changes in work and opportunity. Here, we can only express our appreciation to those with whom we formed close interpersonal ties.

Our colleagues and mentors—including Cynthia Duncan, Phyllis Moen, Marcie Pitt-Catsouphes, and Peter Whalley—offered valuable guidance by directing our attention to the issues that need to be addressed and what to look for. Reviewers Judith Barker, Elizabeth Callaghan, William Canak, Carol Caronna, Marc Dixon, Linda Geller-Schwartz, Heidi Gottfried, Judith Hennessy, Martin Hughes, Arne Kalleberg, Charles Koeber, Kevin Leicht, Joya Misra, Cynthia Negrey, Vincent Roscigno, Gay Seidman, and Patrick Withen provided the sharp criticism that the book needed in its formation. Hillary Gozigian and Stacy Sauppe offered an insightful students'-eye views of the manuscript and helped us bring the manuscript together. Students in our graduate and undergraduate classes at Ithaca College and Cleveland State University, many of whom also are experienced workers in the new economy, raised questions that stimulated our thinking for this book. We also thank the incredibly supportive team at SAGE, including Jerry Westby, Elise Smith, Robin Gold, and our editor, David Repetto.

The Alfred P. Sloan Foundation provided support for the study of job insecurities (B2001–50, Stephen Sweet and Phyllis Moen, co-principal investigators). In addition to Kathleen Christensen at the Sloan Foundation, we extend thanks to Yasamin Diciccio-Miller, Akshay Gupta, and the staff of the Cornell Careers Institute for their contributions to the Couples Managing Change Study.

Finally, we express gratitude to our spouses, Jai and Joyce, who gave us much-appreciated time to devote to this project, who listened to our struggles, and who offered their perspectives and guidance throughout.

Chapter One

Mapping the Contours of Work

Perhaps more than any other quality, the ability to plan, organize, and collectively engage in work sets human beings apart from other species. Work occupies most of our waking hours; it is a crucial part of identities and influences life chances. At the same time, work creates problems in lives and, at its worst, can become a life sentence to grinding toil in jobs that offer few intrinsic rewards and little financial compensation. Our understanding that work can liberate but also enslave—and seeing both possibilities exemplified in the modern economy—inspired us to write this book. We wanted to take stock of work today—to consider the types of work opportunities available, chart how these jobs emerged, and gauge the impact workplace practices have on lives both on and off the job. Beyond this, we wanted to reflect on how work could be organized so that it *makes sense*—so that it provides the resources people need and contributes meaning to their lives.

This chapter begins this discussion by considering the "contours of work." These contours can be thought of as the terrain on which work opportunities are distributed and traversed. The metaphor of contours is useful because, like geographic topographies, work opportunities have been etched into the landscape by long-term historical forces. Some of these forces resulted in profound changes, wherein old ways of working have been abandoned and new methods introduced. This type of radical transformation occurred in the Industrial Revolution of the early 19th century, and some argue that computer and communication technologies are having a similar effect today (e.g., Castells 2000; Piore and Sabel 1984). Other forces, however, shape opportunity landscapes in a more gradual, ongoing, and cumulative process. Gender, with its constantly evolving meanings and practices, is one such force. So are race and social class.

To consider the impact of social forces on work, we introduce here a shorthand distinction that we use throughout this book: the division between the old and new economies. This dichotomy helps us identify the very real changes that occurred in work in the latter part of the 20th century, including the introduction of computer technologies, the expansion of a global economy, shifts in the composition of the workforce, new organizational and managerial paradigms, and other changes that we introduce in the chapters to come. The **old economy** represents the various ways of assigning and structuring work that developed in the wake of the Industrial Revolution through the mid-20th century. It included systems that were built around mass production, gendered divisions of labor, unionized labor, and a variety of other enduring workplace practices. It was also an economy in which the United States was a central and dominating economic force. The concept of a **new economy** is intended to pose the question of whether the nature of work has changed, and if it has, the extent to which these changes are affecting lives on and off the job. Our frame of reference throughout this book is the changing contours of work in the United States, but as they are connected to the redistribution of opportunities in the global economy.

Though we use the term *new economy*, we have come to conclude that many of the present-day contours reflect the way work evolved in the old economy.[1] Those arguing that there has been a "second industrial revolution" often ignore this. There are new jobs, new workers, and new work designs, and these are changing some of the ways work is performed, by whom, and the returns received. But many of the features introduced by the old economy remain. These "old" features are not simply vestiges destined eventually to die out; they are thriving and may be permanent features of the new economy that will continue to develop during the 21st century. Sometimes these old and new features are combined, for example, when old work practices are moved from the developed world to the emerging economies. The jobs may not have changed fundamentally, but the people who are performing them have. In Chapter 2, we consider the issue of production in the old and new economies in greater detail and assess the extent to which the process and organization of working in the new economy has changed and the extent to which it has remained the same.

Our discussion in this chapter is directed to identifying the dominant social forces that shape work opportunity. We organize this discussion by considering three interlocking concerns:

- **Culture:** meaning systems that attach individuals to work, harness their commitments, and direct their efforts.
- **Structure:** opportunities, as well as constraints, that shape what types of jobs can be pursued and by whom, and the returns received.

- **Agency:** people's efforts, whether as individuals or in groups, to direct their own biographies, shape the lives of others, and respond to and sometimes modify the structure and culture of work.

To open this discussion, we consider the lives of five workers laboring in the new economy, and the rewards, strains, and constraints work produces in their lives. As you read these examples of what work is like in the new economy, reflect on the ways current opportunity structures fail to provide the resources needed and think about how work provides meaning but also disrupts lives. The challenge, we argue throughout this book, is considering the best means of bringing culture, structure, and human initiative into harmony. In other words, the goal is to reduce the incompatibilities between how work is arranged and what workers can bring to—and receive from—their work.

Scenes From the New Economy

The experiences of Eileen, Dan, Emily, Jamal, and Chi-Ying reveal how work lives on and off the job are being shaped by the contours of the new economy. All of these cases illustrate that the effects of historical change (in this case the transition to a new economy) can vary depending on its timing with respect to an individual's biography, as well as to his or her gender, class, and race (Elder 1999; Moen 2001). Their lives are unfolding as new opportunities are being introduced and as old opportunities are being dismantled—a dynamic identified as aging on the moving platform of history (Riley and Riley 2000).

Exhibit 1.1	Eileen: A Mother Strives to Mesh a High-Powered Professional Career With Family Demands

Eileen is a busy professional engineer in a large corporation. Early in her career, she worked 10 or 11 hours a day while her husband was in school and continued to work long hours after he graduated. She was a dedicated professional and was on the fast track up the corporate career ladder—until she had her first child. At that point she decided that she should cut back on her hours, something she had not originally planned to do. It was not that she preferred being at home to being at work (she said being home was more "boring"), but she felt someone had to do it and her husband was reluctant to cut back his hours because he had less seniority.

(continued)

(continued)

She persuaded her boss to let her try working 60% of her normal hours on a trial basis. He reluctantly agreed. She worked this new schedule for a few months, but found that her boss still expected more output than she could give. He was unhappy with the arrangement, even though she still was doing a good job in almost everyone else's opinion.

When her company reorganized, Eileen was among the first to be laid off. After losing her job, she realized how much she loved it and the extent to which being a full-time, stay-at-home mom did not suit her interests. She worked hard to find another job at the company, one that required her to return to full-time work. But life has not been easy. Her second child was born with health problems and she and her husband frequently face situations in which someone has to leave work to attend to their son. Eileen would like to get another part-time arrangement but knows that a part-time job where she works means no promotions. So, she continues to work full-time and to wonder why she has to feel bad about being away from work when her child is ill.

Note: Based on *The Time Bind: When Work Becomes Home and Home Becomes Work* (pp. 88–98) by Arlie Russell Hochschild, 1997, New York: Metropolitan Books.

Exhibit 1.2   Dan: An Insecure, Older Worker in a Declining Industry Strives to Salvage a Career

Dan had been an autoworker at GM's Linden Plant in New Jersey for about 10 years. He started working there when he was just 20, thinking it would be just for a while, and he never really enjoyed the work. But the money was good, and he found himself thinking about working his way up to a supervisor's position. While he was working at GM, Dan started a chimney sweep business on the side. He started out small and continued working at both jobs, especially after his wife had kids and quit her job. After a few more years of working two jobs, Dan realized that his GM job was in jeopardy; the union was not as strong as it had been and there were rumors of cutbacks and layoffs. He stayed, though, since it did not seem that layoffs would happen tomorrow and his paycheck was important. Then, the company offered workers a buyout. Dan decided that, rather than risk losing his job, he would take advantage of the offer.

He used the buyout money to build up his chimney sweep business, buy new equipment, and so on. He did really well at first, and initially felt he had made the right decision—he liked the work better and was making a good living. But then the economy took a turn for the worse and his business struggled. He hung on by scrambling to find other odd jobs. He likes what he is doing, but he wonders whether he made the right decision. He knows staying at GM meant doing a job he did not like, and he would have continued to worry about layoffs, but he also knows that his business could fail and that he would have to start over again.

Note: Based on *Farewell to the Factory: Auto Workers in the Late Twentieth Century.* (pp. 3–6) by Ruth Milkman, 1997, Berkeley: University of California Press.

Exhibit 1.3   Emily: A Contract Worker Navigates Insecure Employment

---

Emily is in her late 40s and markets herself as a freelance editor/proofreader. While at one time she was a "regular employee," the job that she previously held vanished when her employer relocated. Her path to becoming a contract worker was not an intended career path. It was a means of rescuing a career that was dislodged.

Emily currently works as an editor with two different employers, a large law firm and small publishing company. She is not considered to be a career employee at either of these companies, does not have a private office, and shares a cramped workspace with another employee. She is not included in many of the office social functions and her level of involvement with other regular employees is quite restricted. Nonetheless, her job as a freelancer provides her considerably more flexibility than her coworkers, and this is something that she values. And, because she is skilled at her work, her compensation is comparable to what the regular employees make.

Because the terms of Emily's employment rest on her employers' interest in hiring her for subsequent work, she keeps watch for opportunities. She believes that her type of job, based on short-term agreements between employees and their employers, is the wave of the future. While she feels secure in knowing how to do her work, she has a constant sense of insecurity in that she does not know what the future holds for her.

Note: Based on *Freelancing Expertise: Contract Professionals in the New Economy* (pp. 2–3) by Debra Osnowitz, 2010, Ithaca New York: ILR Press.

Exhibit 1.4   Jamal: A Disadvantaged Young Worker Strives to Start a Career

---

Jamal is a 22-year-old African American man living in Harlem with his common-law wife. They live in a tiny, untidy one-room apartment where they have little more than heat and a place to sleep. Jamal did not finish high school and eventually got his GED. He is intelligent and ambitious, but was not a dedicated student. He has worked since he was very young, partly because he was determined not to become like his mother, who had a drug addiction.

Despite putting a great deal of effort into finding work, he had very little luck finding a good job. The best position he was able to find was at a fast-food restaurant an hour's commute away, which he found through some of his friends who also work there. The job does not pay well (minimum wage) and he often cannot secure the full 8-hour shift he needs to get by. His boss often sends him home after 5 hours, but sometimes expects him to work longer if needed, or less if business is slack.

Jamal and his wife do not get help from their parents. His mother-in-law is furious with her daughter for marrying Jamal, his father was never in his

(continued)

(continued)

life, and his mother is a drug addict. He is pessimistic about his future and expects to continue working in jobs like the one he has now. But he keeps working, hoping that maybe he will be one of the lucky few who gets a job in a car factory.

Note: Based on *No Shame in My Game: The Working Poor in the Inner City.* (pp. 3–13) by Katherine Newman, 1999, New York: Alfred A. Knopf and the Russell Sage Foundation.

Exhibit 1.5   Chi-Ying: A Daughter Strives to Carve a Career in an Industrializing Economy

Chi-Ying is a young peasant girl from a rural village in Northern China. Two years ago, she migrated to Shenzhen in the south to work at an electronics plant. The work is hard and unpleasant, sometimes exposing her to hazardous fumes. She is required to work 11-hour shifts and sometimes has to work overtime as well, or she might be fired. Being absent leads to fines and punishment, so she comes to work even if she is ill. Her apartment is cramped, she lacks access to clean drinking water, and the air that she breathes is often a choking smog. This is the result of both lax regulation and the proliferation of highly polluting industries, which have transformed Shenzhen from a rural countryside into an urban center within a few short decades.

Chi-Ying's move to Shenzhen was supposed to be temporary. Her parents actually arranged a marriage for her a year ago, but that fell apart when she said she wanted to continue working for a few more years. She had to use some of her own wages to pay the groom's parents back for presents and gifts her family had received. Chi-Ying still wants to keep working at Shenzhen, despite the hardships. She has her own income, which amounts to more than half of what her father earns back home as a farmer.

Although she has little time for leisure, she does have the opportunity to buy a few things for herself and to experience life outside her village. She has married a boy she met in Shenzhen who comes from her part of the country, and she knows she eventually will return. But she is pleased that, unlike her grandmother, she has seen life in the city and has had a job of her own.

Note: Based on *Gender and the South China Miracle* (pp. 3–9) by Ching Kwan Lee, 1998, Berkeley: University of California Press.

For Eileen and Chi-Ying, economic changes have opened new opportunities, but they coexist with enduring sets of expectations about what mothers and daughters should provide to their spouses, parents, and children. These cultural orientations seem more appropriate to another era and lag behind what these workers ideally want for themselves and, in many instances, what they can provide for others. For Dan, the moving platform of history has introduced new career tensions because the types of jobs he performed during his entire career are becoming more difficult to find. Consequently, his income is in jeopardy and he faces the daunting challenge of fitting himself into new lines of work that he has little experience performing. Emily is one of a new breed of contract workers, and she is figuring out how to navigate careers with employers who are increasingly inclined to hire workers for short-term task assignments rather than as long-term career employees. Jamal is doing the "right thing," working hard and trying to get ahead, but his opportunities are limited by the fact that good jobs that only require a limited education are disappearing in the new economy.

The careers of these workers are influenced by demands and social ties off the job. All these workers are making career decisions in the context of their linkages to others. In some circumstances, children hold sway, whereas in other cases, it is the needs of spouses, aging parents, or both (Neal and Hammer 2006; Sweet and Moen 2006). These life stage circumstances play an important role in shaping worker behavior, expectations, and needs. How people respond to these circumstances is heavily influenced by cultural scripts (e.g., assumptions about what parents should provide for their children) and the availability of resources (which varies from person to person, group to group). And beyond family ties, the contexts of neighborhoods and communities influence one's ability to find work, the resources to prepare for work, and the security to engage in work (Bookman 2004; Sampson, Morenoff, and Gannon-Rowley 2002; Sweet, Swisher, and Moen 2005; Swisher, Sweet, and Moen 2004; Voydanoff 2007).

Culture and Work

Eileen presents an interesting case to consider because she is a worker who could potentially leave her job to tend to her children. She feels guilt about not doing so, but chooses to stay in the labor force and retains ambivalent feelings about her choices. Understanding why workers like Eileen place such a strong emphasis on their work roles requires considering culture— the meaning systems that surround work and shape identities in respect to it.

Most classic theories of work embrace cultural perspectives that view labor, in and of itself, as a noble endeavor. Karl Marx (1964), for example, argued that work is what distinguishes humans from other species, and he highlighted how it enables people to transform their environments to suit human interests. Sigmund Freud ([1929] 1961) argued that work is a socially accepted means by which humans can direct their sublimated sexual energies. As such, he saw work as a means of achieving satisfaction when fulfillment in other parts of life is lacking or prohibited. Émile Durkheim ([1895] 1964) offered a different thesis, that work and the complex division of labor in society offered a means to create social cohesion. All these perspectives have in common the assumption that work has the potential to cement social bonds and advance the development of civilization.

But has work always been embraced by cultures as playing a central role in people's lives and the workings of societies? Anthropological and historical studies suggest otherwise. In many cultures, work is defined as the means for day-to-day survival. Subsistence economies operate on the basis of cultural assumptions that work is primarily a means to an end, so that once individuals have enough food and shelter, labor is expected to cease. Such an orientation to work in today's American culture would indicate a moral weakness and be perceived as a threat to social order. But from the point of view of many other cultures, our embrace of work could be considered pathological. If one can obtain enough to eat and gain sufficient shelter by working a few hours a day, so be it. Why should a hunter set out in search of game if the supply of food is adequate (Brody 2002; Sahlins 1972)?[2] And even within geographic regions such as modern day Western Europe, expectations regarding the age at which individuals are expected to embark on careers, or to retire, vary remarkably. These varied life-course scripts result in some cultures expecting 10 years (or more) of additional attachment to the labor force than others, with policies that match these expectations (Sweet 2009).

One important cultural question concerns why work plays such a central role in some societies but not in others. Part of the answer, according to Max Weber ([1905] 1998), is that the societies that were in the forefront of the Industrial Revolution had been swayed by changing religious doctrines. These religious beliefs, particularly those that underpinned the Protestant Reformation, created anxieties about one's fate in the afterlife. In response, Western European and American culture advanced the value of the work ethic, a belief that work is not something people simply do, but is a God-given purpose in life. Devoting oneself to work and doing a good job were considered to be ways of demonstrating to oneself that a life of virtue reflects grace. And as members of these societies embraced the idea that work is "a calling," they applied themselves to their jobs with greater vigor, creating wealth and affirming to themselves and others that God was looking favorably on their actions.

Although many now question Weber's thesis that the Protestant Reformation was responsible for the emergence of capitalism, the centrality of the work ethic to the development of Western society is widely accepted.

So deeply is it ingrained in contemporary American culture that nearly three-quarters of Americans report that they would continue to work, even if they had enough money to live as comfortably as they would like for the rest of their lives.[3] Americans work to affirm to themselves and others that they are virtuous, moral individuals, good people who deserve respect (Shih 2004). Conversely, those who choose not to work, or workers like Jamal who are unsuccessful in securing a job, are looked down upon and stigmatized. In American society, to be without work is to be socially suspect and unworthy of trust (Katz 1996; Liebow 1967).

The work ethic defines labor as a virtue, but it also has pathological dimensions. The cultural embrace of work may be akin to the flame that attracts the moth. It is telling that many who can afford to work less, and who have the opportunities to do so, choose not to (Hochschild 1997). Psychologists call these individuals "workaholics," (Machlowitz 1980), but as we discuss later in this book, many of those driven to work long hours do so because they are driven by organizational cultures that bestow rewards on those who live, breathe, and eat their jobs. The suspicion cast on those who do not hold jobs has created pressures to force work on those who get little benefit from it. Consider that welfare reform legislation, passed in the mid-1990s, requires even very poor mothers of young children to work to receive welfare assistance. This requirement defines mothering as "not work" (a concept we return to later) and accepts the fact that many of the affected mothers remain in poverty even after they are employed.

Thorstein Veblen ([1899] 1994) in *The Theory of the Leisure Class* observed that attitudes to work are bound up with materialistic values held in American culture. Markers of status include luxury autos, large homes, and expensive clothing. All of these markers of success are conspicuously consumed, put on display to be seen and admired, and set standards for others to follow. By the mid-20th century, the drive to purchase social status had permeated American society, compelling workers to labor hard "to keep up with the Joneses" and their neighbors' latest purchases (Riesman, Glazer, and Reuel 2001). Contemporary American workers engage in the same status game that emerged in the late 19th century, but with new commodities (e.g., iPads, BMWs, and flat screen TVs). Their competition now expands beyond their neighborhoods, as they are literally saturated with media images of success and have developed numerous ways to accumulate debt (home equity loans, student loans, and credit cards) (Gergen 1991).[4] The result, some have argued, is "affluenza," the compulsion to purchase and spend beyond one's means (de Graaf, Wann, and Naylor 2001). For some members of the new economy, work has become the means to manage spiraling debts incurred while striving to keep up with others who are spending beyond their means as well (Schor 1998).

Culture also shapes the attitudes workers and employers have toward each other. One means by which it does this is by constructing social divisions and setting group boundaries. Racial and gendered divisions, for example, are based on assumptions that different social groups possess different capabilities. In turn, these beliefs contribute to the formation of self-fulfilling prophecies. Whether these differences were originally real is immaterial; as the early 20th century American sociologist W. I. Thomas noted, what people believe is real often becomes real in its consequences (Thomas and Thomas 1928). As we discuss later in this book, these self-fulfilling prophecies about gender and race shape social networks, influence access to resources, and funnel people into different lines of work.

Culture even extends into the design and management of jobs and technologies. Consider, for example, the enduring legacy of scientific management. Frederick Winslow Taylor introduced this managerial philosophy (also known as Taylorism) at the beginning of the 20th century to increase the productivity of workers laboring in factories. He advocated the benefits of redesigning work to wrest control from workers and place it in the hands of management. His *Principles of Scientific Management* ([1911] 1964) argued for the separation of "thought" from "execution" to establish clear divisions between managers (whose job was to think and design) and workers (whose job was to carry out managers' instructions). He used time-motion studies to decompose production jobs into the simplest component tasks in order to increase worker speed and accuracy. And managers' jobs were redefined to absorb worker skills into the machines and organization and to keep the flow of knowledge going in one direction—from the shop floor into managers' hands. The result was the creation of legions of deskilled jobs, the dissolution of many craft skills, and a decline in the individual worker's ability to control the conditions and rewards of work (Braverman 1974; Noble 1979; Pietrykowski 1999). It also fostered distrust and hostility between workers and their bosses (Montgomery 1979).

Why did Taylor advocate this way of organizing work, given its obvious negative consequences for the quality of work life and its negative effects on labor-management relations? In part, it was a response to something real—the fact that workers often did not work as hard as they could. His experiences had taught him that they did not show up to work consistently, took long breaks, and worked at a more leisurely pace than owners desired. His interpretation of this behavior, however, was culture-bound. Taylor interpreted workers' behavior not as a rational, class-based resistance to employers but as an irrational unwillingness to work in the right way. Taylor, like many Americans of his time, was embracing a cultural denial that class divisions within the workplace existed. His solutions also reflected the culture in which he was living. He advocated a reorganization of the workplace based on scientific methods, something that resonated tremendously in a society where science had come to be seen as the solution to many human problems. And he depicted the worker as essentially unintelligent and easily manipulated; Taylor was fond of using an example involving a worker named Schmidt (whom he described as "oxlike"), whom he persuaded to adopt his new system through a combination of simple-minded arguments and limited incentives. This, too, was typical of American culture at that time; many Americans believed that members of the lower classes, immigrants, and others at the bottom of society were inferior in various ways (including intelligence) to the more successful members of society. Taylor's ideas also reflected an abiding cultural belief in the correctness of capitalism, particularly the proposition that it is natural that some should be owners and others laborers, that the efforts of those at the top were more important and valuable, and that an extremely unequal distribution of the fruits of labor was not just defensible but actually desirable (Callahan 1962; Nelson 1980).

Exhibit 1.6   The Film *Modern Times* Offered a Poignant Illustration of the Alienating Nature of Work in Factory Jobs in the Old Economy

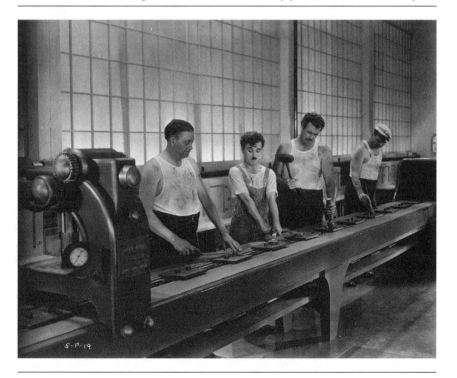

Source: Copyright © by Getty Images. Reprinted with permission.

The legacy of managerial philosophies—in this case, scientific management—highlights how culture and social structure intersect. Managerial perspectives that embraced the proposition that workers are indolent and should not be trusted are directly responsible for the creation of many of the alienating, low-wage "McJobs" present in America today. These philosophies initiated the development and application of assembly lines, promoted the acceptance of the idea that some people should be paid to think and others to labor, and fostered divisions between "white-collar" and "blue-collar" jobs. Dan performed blue-collar work during the bulk of his career, and these approaches to organizing work are directly responsible for shaping Jamal's tasks in the fast-food restaurant.

These examples of how culture shaped workplaces in the past suggest interesting questions about culture's role in carving out the contours of the new economy. Have cultural attitudes about the role of work changed, and if so, have workplaces changed along with them? How long are people working and why do they work so much? Have Americans begun to abandon long-standing (Taylorist) cultural assumptions about the proper way to organize work, or do we continue to construct workplaces on the assumption that workers are lazy, ignorant, and not to be trusted? To what extent are perceived divisions between the members of society continuing to deprive some people of access to opportunity? We address these concerns in the chapters that follow.

Structure and Work

In contrast to culture and the way it creates meaning systems that orient people to work, social structure imposes enduring patterns of social organization that determine what kinds of jobs are available, who gets which jobs, how earnings are distributed, how organizational rules are structured, and how laws are formulated. Social structure does not exist independently of culture. Often, social structure reflects cultural attitudes, because people tend to create institutions that are consistent with their beliefs. And it also can be in conflict with aspects of culture, creating tensions and contradictions with which individuals and societies must grapple. Consider how the structural reality of unemployment creates particularly difficult problems in a society where work is valued or structured as a mandatory part of citizenship. Throughout this book, we discuss various aspects of social structure, the access to different types of work, the division of labor, the social organization of workplaces, and legal and political arrangements that influence the design of jobs and the terms of employment. Here, we simply illustrate how social structure affects individuals' experience of work by considering how a few aspects of social structure—social class, job markets, and labor force demographics—may be affecting opportunities and workplace practices in the new economy.

Class Structures

Witnessing the changes wrought by industrialization, Marx ([1867] 1970) focused his sociological analysis of work on **class structures,** the socioeconomic divisions between different segments of the workforce. In his classic analysis of the industrial capitalist economy, he argued that employers' profits depend on the effort put forth by employees, which created incentives to limit wages and to push workers to labor as hard as possible. He also observed that the efforts of workers created far greater wealth for employers than it did for employees. Considering these class relations, Marx argued that the tendency for work under capitalism would be toward the creation of a polarized class structure, comprising a disenfranchised working class (the proletariat) and an affluent owner class (the bourgeoisie).

The contemporary class structure of the United States is more complex than the polarized structure Marx envisioned that it would be. Although workers and capitalists exist, large portions of the workforce seem to fit into neither category. For example, numerous professional and managerial workers have substantial education, some (or even considerable) workplace authority, and higher salaries than the typical frontline worker. Yet, it is difficult to describe them as captains of industry or members of the dominant class, given that they are not in charge and work for someone else (who has the ability to fire them). Sociologists have argued long and hard about how to describe these intermediate class positions. One sociologist described such workers as occupying "contradictory class locations," combining elements of the classes above and below them (Wright 1985).

Though the precise shape of the class structure of capitalist societies is a matter for dispute, what is not disputed is that class matters. For example, class affects people's access to work opportunities, although the precise way in which it does so has changed over time. Before the Industrial Revolution, most children inherited their line of work from their parents through a process known as **ascription.** Farmers' children tended to become farmers themselves, and craft workers would often learn their trade from their fathers. Women's roles were largely ascribed as well. One's occupation was to a great extent one of the things one inherited from one's parents; the cross-generational effects of class were obvious and straightforward. With industrialization, however, the range of jobs expanded profoundly, many new occupations were introduced, and other occupations became less common or actually disappeared. As a result of these changing opportunity structures, fewer children could follow in their parents' footsteps or inherit occupations from the previous generation. By the late 19th century, geographic mobility and social mobility became more common, as children ventured further from their home communities to find work (Thernstrom 1980). Class still mattered, however, because it affected one's access to resources such as education, skills, and connections that determined access to work in an economy where jobs no longer were inherited.

The existence of class also affects the structure of workplaces. Marx felt that the antagonism between labor and capital inevitably produced antagonism at work and led to the development of hierarchical, top-down managerial structures designed to control workers and ensure that the interests of employers predominated. Although the polarized workplaces envisaged by Marx may not be the dominant organizational form, managerial efforts to control labor reflect a strong desire to respond to class antagonisms. They reflect the reality that the workplace is a zone of contested terrain, one in which class conflicts take place, with each side using the weapons at its disposal—including layoffs, speedups, technology, strikes, and even sabotage (Edwards 1979; Montgomery 1979).

Throughout this book, we argue that social class remains one of the most powerful forces shaping employment opportunities and access to resources in the new economy. We examine how a changing economy has altered the reality of class and the extent to which changes in class structure have led to a fundamental restructuring of workplaces away from the familiar patterns of industrial America. We also examine how gender and race matter and how they interact with class to shape complex, contemporary structures of opportunity and workplaces. We focus on social class most directly in Chapter 3, but throughout we emphasize that other social markers (such as gender, race, or age) intersect with class in important ways.

Job Markets and Job Demands

Jamal's and Dan's problems involve not simply finding work, but rather finding work that pays a reasonable income. Eileen, on the other hand, possesses highly marketable skills and can command a handsome salary. For her, the problem is securing a job that is designed to correspond with what she can bring to the job. But for Emily the contract worker, the problem is locating new opportunities that enable her to navigate from one job to the next. All four of these workers have concerns that are structural in nature and involve the way opportunities are configured. All have to adapt themselves to the existing range of jobs and the prevailing ways in which jobs are organized. Their personal problems reflect the fact that workers—especially those laboring in times of economic change—face challenges in locating and adapting themselves to opportunities. For many workers, and those left involuntarily out of the labor force, the most recent economic recession has significantly compounded these problems.

The Industrial Revolution of the early 19th century was clearly a watershed, one that profoundly reshaped the types of jobs available to workers. The most obvious consequence of industrialization was that far fewer people were employed in agriculture and many more were employed in factory work. However, the changes were not limited to the shift from agriculture to industry. Traditional occupations outside agriculture were also transformed, as new technologies and new ways of organizing work pushed older approaches aside. For example, the mechanization of weaving during the Industrial Revolution completely removed this work from the home and introduced new skills that fit factory labor. Similar stories can be told about many other traditional occupations, including hat making, shoe production, tanning, and tinsmithing (Thompson 1963).

Is the range of employment opportunities available to American workers changing again? It certainly seems that way. Some jobs that used to be plentiful in America have virtually disappeared, and the skills needed to obtain jobs are changing as well. In the old economy, for example, it was common for children to follow their parents into the mill or factory and receive good wages for performing jobs that required little education. Today, few young people aspire to become steelworkers or factory operatives, largely because many of these jobs have disappeared. As steel mills and factories closed in the 1970s and 1980s, the impact reverberated throughout industry-dependent "rust belt" communities, forcing their residents to rethink long-standing beliefs about jobs, futures, and how one makes a living (Bartlett and Steele 1992; Bluestone and Harrison 1982; Buss and Redburn 1987).

One way of considering the changing opportunity landscape is to consider the process of **creative destruction**, a phrase introduced by the economist Joseph Schumpeter (1989) to describe the tendency for old methods of production to be replaced by newer, more efficient approaches. In some cases, new technologies make old needs obsolete, as when the automobile extinguished the need for buggy whips. In other instances, technological innovation can replace workers with machines, as was the case with cigarette rollers (Bell 1973). New methods of organizing work can be used also to reduce production costs, for instance, by moving jobs to locales where labor costs are lower (Cowie 2001). And in the case of computers, technologies have not only replaced workers, but also introduced entirely new markets and jobs.

The drive to create ever more efficient and profitable enterprises is influencing the distribution of work opportunities around the world. Production now occurs on a global scale, and the forces that disperse work to far-flung locations such as Indonesia (where athletic shoes are assembled) and Vietnam (clothing) shape the life chances of workers both at home and abroad. Understanding the reasons why work is being dispersed, and the impact on workers' lives at home and abroad, is essential to revealing the trajectory of work and opportunity in the new economy. Throughout the 20th century, the United States held a dominant position in the global economy. But in the new economy, jobs previously held by Americans such as Dan are increasingly being exported to countries such as China and India and being performed by workers like Chi-Ying. Is this resulting in deteriorated or enhanced opportunities? And for which workers?

Changing employment opportunities also have redefined what skills are needed, reshaped job demands, and introduced new rewards. They also impose new burdens on workers' lives. Consider the large number of jobs available in various kinds of **interactive service work** that emerged in the latter part of the 20th century. These jobs require a different type of work than that performed in the factory, in that the employees typically do not manufacture anything. Workers such as teachers, therapists, or servers provide a service for someone else with whom they are in direct contact. Sociologists have noted that this kind of work places different demands on the worker (Mills [1951] 2002; Paules 1991). He or she must learn interaction skills—how to make others feel comfortable, how to produce the desired kind of social setting, how to deal with various kinds of difficult social situations—*because the interaction is a significant part of the product being sold.* The work of airline flight attendants offers a compelling illustration because they are trained to make customers feel safe in the rigid and sometimes frightening environment of an airplane. To do this, these workers are coached on techniques to change their internal emotional states to generate the display of warmth or sex appeal required by their employers. As a consequence, however, these types of workers are especially prone to experiencing emotional numbness or burnout (Hochschild 1983).

New jobs demand new sets of skills, but new technologies and organizational systems also are transforming many familiar jobs. A secretary's job, for example, is quite different than it once was because computers have eliminated aspects of the old job (repetitive typing) and created new ones (basic graphic design, data analysis, electronic communication). Bank tellers once were simply clerical workers who processed clients' financial transactions. Now, however, computerized information systems provide tellers with information about clients' financial positions and prompt tellers to sell various products to the client, all while a close electronic eye monitors what the worker is doing (Smith 1990). Even traditional manual labor is affected. For example, production workers who used to rely on their senses of touch and smell as guides now work in clean settings and operate sophis- ticated computerized systems that make some of their old ways of working obsolete (Noble 1979; Shaiken 1984; Vallas and Beck 1996; Zuboff 1988).

Finally, job opportunities may be less rigidly tied to space and time than they were in the old economy. Today, many workers have opportunities to telecommute and work from home offices. The economy operates 24/7, introducing the prospects of working alternate shifts and reconfiguring work around family lives. This may open opportunities to liberate workers from the traditional 9 to 5 grind and introduce new flexible schedules that more harmoniously mesh work with life—a work arrangement that Eileen's boss was reluctant to accommodate. However, it may also open prospects that work will intrude on lives in ways not possible in the old economy. Understanding the impact of these new structural configurations is essential to charting the contours of work in the new economy.

Demography and the New Labor Force

The composition of the workforce is undergoing change as well. As a result, sociological analysis of work requires consideration of **demography** and of how the composition of a society affects the placement of workers into jobs and the distribution of opportunities to prepare for and obtain work (Farnesworth-Riche 2006). The paid labor force is quite different today than it was in the mid-20th century or earlier. It contains a far higher percentage of women, and its racial and ethnic makeup is different. We devote two chapters of this book specifically to the issues of gender (Chapter 6) and race and ethnicity (Chapter 7); here we introduce the importance of demographic forces by considering how age structures affect the availability of jobs, the availability of workers, the need to work, and the returns received from work.

The U.S. labor force, along with those of many other developed soci- eties, is aging. Americans today can expect to live 12 years longer than could those alive in 1940, and 26 years longer than those who were alive in 1900.[5] Workers are living longer, and they are healthier when they reach ages that used to be considered "old." This presents new opportunities, as well as new challenges, to American workers and their employers, such as the approach to dealing with retirement. Should workers continue to stop working at 65 if they will live for many years after that? If people are living longer and staying healthy longer, perhaps work careers should be lengthened. However, older workers generally do not want jobs that demand heavy schedules. More common are desires to enter into second or third careers and to pursue work situations that focus less on earning money (although for many that remains important) and more on satisfying

Exhibit 1.7   Age Distributions in the United States: 1940 and 2010

Source: U.S. Census Bureau.

creative desires or making a difference in the lives of others. Unfortunately, most employers do not offer "bridge jobs" that accommodate the possibility of the types of scaled-back employment that fit the skills and interests of these workers (Hutchens and Dentinger 2003; Moen and Sweet 2004; Moen, Sweet, and Swisher 2005).

The changing age structure of the workforce presents challenges to society as a whole, not just to employers. Exhibit 1.7 shows how the age structure of the United States has changed from 1940 to 2010. Note that in 1940 the age structure of the United States resembled a pyramid, with most of the population in the younger age groups, with a steady attrition as one approached old age. Only a relatively small group lived beyond age 70. In contrast, in 2010 the age pyramid looks more like a skyscraper, albeit with a bulge in the middle. This bulge is the baby boom generation, a birth cohort that is steadily aging its way into retirement years. A key structural question concerns how an aging society will provide economic support for the growing numbers of older people. Will they be required to work? Or will society continue to provide postemployment pensions for them? And, if the latter, how will that expense be financed? The Social Security system, the most important source of retirement income for many Americans, is funded through taxes on currently employed workers. Those taxes become part of the general pool of Social Security revenue, which provides pensions to those who have retired. Some policy makers are concerned that if the pool of retired workers becomes larger and the pool of employed workers becomes smaller, the revenues available to fund the system will be squeezed (Weller and Wolff 2005). There is much controversy about whether this should be called a "crisis," but there is general agreement that ways need to be found to ensure that adequate revenues will be available for the growing population of retired workers.

Demographic factors such as age, gender, and race affect virtually all aspects of the economy and workplace. Demographics play a role at the organizational level, as the experiences of ethnic minorities and women are commonly shaped by their scarcity at the top levels of organizational hierarchies. They are critically important at the community level, as neighborhoods that lack job opportunities hinder the socialization of children into the types of workers needed in the new economy. We return to the critical issues of aging, gender, race, education, and immigration throughout this book.

Agency and Careers

Sociologists are often accused of arguing that people are simply "pawns" or "cultural dopes" of the larger social structural and cultural contexts in which their lives are lived. The depiction of individuals as victims of external forces ignores agentic capacities—their ability to direct their own lives and those of others (Emirbayer and Mische 1998; Garfinkel 1967; Wrong 1961). All of the workers we considered made choices. Jamal got married at a young age and dropped out of high school, Eileen elected to have two kids and pursue a high-powered career, Dan took the initiative to start his own business, Emily keeps her eyes and ears open for new work, and Chi-Ying chose to move from her village to the city. These observations highlight the ways in which different people direct their life courses and how access to different resources and constraints shapes how lives are constructed over time (Elder 1998; Moen 2001; Sweet and Moen 2006). The life-course perspective is essential to understanding the contours of the new economy because it focuses on careers—the patterns of entry, exit, and movement between jobs.

Agency, of course, depends partly on resources. People with unlimited resources at their disposal are in a far better position to design their own lives than are those who have few resources. The new economy may be creating a context that is expanding the control individuals have to direct their life courses, in essence making lives less scripted than in the old economy (MacMillan 2005). Many old structural barriers have been removed (such as segregation laws), and so have the cultural barriers that funneled women and ethnic minorities into restricted ranges of occupations. Before the enactment of civil rights legislation and the women's movement, the prospects that women like Condoleeza Rice, Hillary Clinton, or Sonya Sotomayor could move into positions of power were slim to nil. Today, one can quickly generate a sizable list of minority group members and women who have moved into professions in which they once had been entirely absent. Still, ample evidence indicates that women and minorities are at distinct disadvantages in securing many types of jobs (Grusky and Charles 2004; Reskin, McBrier, and Kmec 1999). Whether the new economy is fundamentally altering the possibilities for people to shape their own biographies is one of the central questions posed in this book.

Agency also plays a critical role in shaping the way work is performed. Numerous ethnographic studies reveal that workers are not simply passive recipients of culture and structure; they use personal initiative to influence how their jobs are performed and the returns they receive from work (Darrah 2006; Montgomery 1979; Richardson 2006; Roy 1955; Tulin 1984). To illustrate agency at work, consider Michael Burawoy's (1979) observations of production workers in the machining industry. These workers' jobs were regulated by quotas, wherein they had to make a specified number of parts to earn their base pay. But when they surpassed those quotas, they could "make out" and earn additional money. In one respect, this system was rigged by management to increase productivity. However, Burawoy observed that the machinists invented a variety of tricks to game this system. For example, they would keep quiet about the easy jobs in which quotas were underestimated, and complain incessantly about the impossibility of meeting quotas on virtually all other jobs. They would bribe supervisors to get the easiest jobs and curry favor with coworkers to provide the stock needed to get their jobs rolling. When given an easy quota, workers overproduced and then hid their "kitties" that could be turned in for extra compensation at a later date. In sum, these machinists showed that when workers are confronted by cultural and structural arrangements, they also engage in **strategic action** to influence how these arrangements affect their lives (Moen and Wethington 1992; Sweet and Moen 2006). Underpinning much of the research on agency is the question of equity and how it is socially negotiated. One wonders, if the economy has changed significantly, have the strategies workers use to assert their will changed as well?

Finally, it should be added that agency also operates at a collective level. Workers make efforts to carve out work lives for themselves, but they also collaborate to reshape the contours of work and create more satisfactory work opportunities for others. An obvious example is that workers band together in organizations such as unions or professional associations that use the strength of numbers to press for needed changes. Union publicity materials that describe unions as the "people who brought you the weekend" remind us that collective action obtained the taken-for-granted days off workers now enjoy. Similarly, the professional associations formed by doctors, lawyers, and others help protect those workers from competition, define what are acceptable (and unacceptable) professional practices, and generally shape the conditions under which those types of work are performed. Throughout this book, and particularly in the concluding chapter, we examine how collective action has shaped workplaces in the past and how it might do so in the future. Is the new economy making certain forms of collective action by workers obsolete? Is it creating openings and needs for new kinds of collective action? What are the key issues around which workers band together to effect change?

Conclusion

In this first chapter, we focused on the ways that sociological perspectives reshape the consideration of work. Although work is commonly considered a means to obtain a paycheck, we argue that it is much more than that. The design of work corresponds with cultural templates that guide workers to their jobs and script social roles. Workers live within social structures that allocate opportunities and construct barriers that block access to meaningful employment. And within these contexts, workers have responded both individually and collectively to manage their responsibilities and reshape society.

The stresses experienced by workers like Eileen, Chi-Ying, Dan, Emily and Jamal are probably familiar to many readers of this book. Because of the instability of jobs, changing opportunity structures, the challenges of meshing work with family, and the challenges of finding good work, many workers find themselves struggling in the new economy. One of the great contributions of sociology is its capacity to reframe these types of personal problems as being public issues (Mills 1959). In the chapters that follow, we consider the extent to which work opportunities are changing, and the impact these changes have on lives on and off the job. Our focus, throughout, is to identify stress points, opportunity gaps, the ways in which workers adapt to these strains, and what can be done to close the chasms that separate workers from fulfilling jobs and reasonable conditions of employment.

Notes

1. Of course, these are not the only phrases used. Others use the term *Fordism* to describe the old economy, and depending on the political slant of the analysis, *post-Fordism* and *flexible specialization* are used to describe the new economy, as are *knowledge economy, global economy,* and postindustrial economy (Bell, 1973; Hirst & Zeitlin, 1991; Piore & Sabel, 1984).

2. It is worth emphasizing that describing societies such as these as "poor" is misleading. Though they lack the variety of possessions contemporary Americans enjoy, their members often live healthy and fulfilling lives.

3. Authors' analysis of the General Social Surveys. Retrieved from the General Social Survey website: www3.norc.org/GSS+Website/.

4. Approximately one-third of American families rent their homes, one-quarter live at or near the poverty level, and nearly one-half will experience divorce (Coontz 1992). These facts are seldom represented in television's portrayals of the "typical" American family.

5. American men now live, on average, to be 76 years old, and American women have a life expectancy of 81 years.

Chapter Two

New Products, New Ways of Working, and the New Economy

One of the most popular themes in discussions of work is the idea that recent changes in work constitute the equivalent of a second industrial revolution. Consider, for example, the impact computers have had on the ways jobs are performed and designed. Computers enable workers to correspond at great distances, telecommute from home, and access a wide array of information. These "smart machines" have absorbed many workers' jobs and replaced human hands with robotic pincers that move with exacting precision. Computers also have spawned new markets for software and hardware, creating new jobs requiring new skills. Their reach spans the world, enabling near-instantaneous transmission of information, as well as the coordination of complex trade relationships that link companies with one another in global webs. It is hard not to conclude that computers have sparked revolutionary changes—not only in what is being produced and how jobs are designed, but also in the geographic distribution of work. What impact do these types of changes have on current and future generations of workers?

The use of the concept of a "new economy" (or alternate terms such as *global economy*) is widely accepted as a shorthand way of saying that work today is remarkably different than it was in the recent past. But in this chapter, we open this assumption to debate. If there is a new economy, what are its distinguishing characteristics? We argue that jobs *have* changed in profound ways. There are new technologies, organizational designs, industries, and markets. The economy has become increasingly international. These changes have introduced the need to develop new skills to fit changing opportunity structures. But what is equally true is that many aspects of the "old economy," including the design of jobs to require limited skill, have either survived or been reproduced in new forms. After all, for every successful computer programmer who works at a company like Microsoft, one can find three poorly paid workers laboring on hamburger assembly lines at companies like McDonald's.[1] Understanding the new, the old, and the old in the new is the key to understanding the diverse needs and experiences of today's workforce.

In this chapter we consider some of the major changes said to characterize work in the new economy, including the decline of mass production and manufacturing work, new skill requirements, the emergence of new cultures of control, the gradual decline of organized labor, the rise of flexible work arrangements, and globalization. In each case, we argue that there have been significant changes, but also that there are persistent features that reflect the perpetuation of the old economy within the new.

A Postindustrial Society?

One of the earliest forecasts of an emergent new economy came from the sociologist Daniel Bell (1973), who argued in the early 1970s that America was entering a "postindustrial" era, in which the manufacturing-centered economy of the past was being replaced by an economy directed toward the provision of services. Bell was among the first to note something that subsequently became obvious to most Americans, particularly those located in the so-called rust belt of the industrial Midwest—employment opportunities had shifted away from manufacturing to other sectors of the economy.

Exhibit 2.1 shows that in 1940 the number of employees working in the manufacturing sector in America was more than double that in any other sector of the economy, accounting for over one-third of all employment. Until 1989, the manufacturing sector remained the dominant employer. But as the population of the United States grew during the latter part of the 20th century, manufacturing employment did not. Today, instead of employing one in three workers, as it did in the mid-20th century, manufacturing enterprises only employ one in 10 workers.

There are various explanations for this trend. Some argue that nearly all low-skill, low-wage manufacturing work is being funneled to developing economies, while the advanced economy of the United States focuses on knowledge work and services (Bell 1973; Fröbel, Heinrichs, and Kreye 1982). However, it is also possible to argue that this simply reflects something "old"—the continued effort of employers to find the least expensive ways to produce goods (Cowie 2001). From this point of view,

Exhibit 2.1   Trends in Employment in 12 Major Sectors: United States 1940–2010

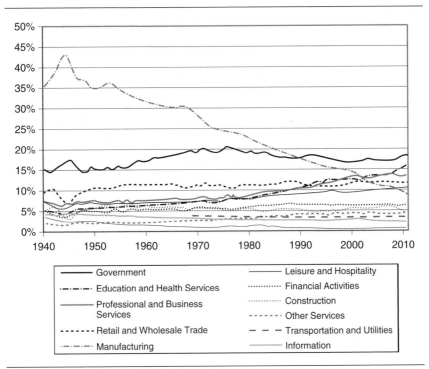

Government
Education and Health Services
Professional and Business Services
Retail and Wholesale Trade
Manufacturing
Leisure and Hospitality
Financial Activities
Construction
Other Services
Transportation and Utilities
Information

Source: Bureau of Labor Statistics.

manufacturing remains central to the economy; however, it now takes place on a global scale, rather than on a national one. Yet another interpretation emphasizes that the United States is unusual—the decline of manufacturing employment is more pronounced here than elsewhere. Rather than reflecting a long-term, general trend away from manufacturing, the U.S. pattern may reflect a choice by American employers to seek low-wage sites for manufacturing rather than invest more heavily in improving techniques at home (Appelbaum and Batt 1992). All of these processes have played a role in shaping opportunity in the new economy.

The number of manufacturing jobs has declined in the United States and other economically advanced countries, but should we conclude that we are truly postindustrial? In the new economy, manufacturing enterprises continue to employ more than 11 million American workers. Although manufacturing employs a smaller percentage of Americans than it once did, it remains a major force in the economy and creates demand for the products and services generated in other parts of the economy (Glyn 2005). It is not at all clear that manufacturing employment is in an inevitable long-term decline to the point where it will disappear entirely. Rather, it remains an important, but less dominant part of what is now a more diversified economy.

Finally, it should be recognized that the practices of the old industrial economy have been woven into the design of many jobs central to the new economy. The fact that manufacturing opportunities have stagnated in the United States does not mean that the *ways of working* that developed in the old economy are on a path to disappearing as a result. The degree to which the old industrial economy has or has not been replaced can be examined by considering the extent to which mass production is disappearing.

The End of Mass Production?

It is generally agreed that economic activity in the old economy centered on the production of manufactured goods (e.g., automobiles, steel, chemicals, appliances) in large quantities for mass markets. The Ford Model T is the classic example of what the American manufacturing economy produced— an affordable and highly standardized car, mass-produced by American workers in a central factory location (Chandler 1990). Coordinating hundreds (and sometimes thousands) of workers at a single location meant that employers such as Henry Ford had to develop bureaucratic management systems, complete with rigid job definitions, rules of conduct, and productivity expectations (Edwards 1979).

The dominant managerial approach of the time was to follow the practices of scientific management, which encouraged the replacement of skilled workers with cheaper, more dispensable low-skill workers, while removing discretion from the shop floor and placing it in the hands of management (Braverman 1974). Finding ways to enhance productivity through job simplification, replacing people with technology, and improving managerial control over what was happening in the workplace were all central to this approach. The assembly-line epitomized this philosophy, a combination of technology and organization that harnessed workers to labor at repetitive, simple tasks (Hounshell 1984; Meyer 1981; Noble 1984; Sabel and Zeitlin 1985). To appreciate how this affected the performance of work, consider the difference between making automobiles using highly skilled workers (as many of the first automobiles were made) and the assembly line methods pioneered by Henry Ford. Instead of relying on a skilled (and hard-to-replace) craft worker who controlled how the work was done, the assembly line created jobs that required very little training, involved relatively simple, repetitive operations, and dictated to the worker how the job should be performed. Perhaps most importantly, the worker lost control over the pace of work, as the assembly line pushed work forward at a speed primarily determined by management. The result was the creation of legions of deskilled jobs, the dissolution of many craft skills, and a decline in the worker's ability to control the conditions and rewards of work (Braverman 1974; Pietrykowski 1999).

This approach was enormously successful and formed the basis for the growth of the giant American manufacturing enterprises (e.g., Ford, General Motors, U.S. Steel) that dominated the American economic landscape and symbolized American economic power worldwide. This approach also fostered distrust and hostility between workers and their bosses, who developed an "us versus them" mentality, in which each side saw the other as having interests fundamentally opposed to its own. Thus, as managers tried various tricks to speed up work, those laboring on the front lines developed alternate approaches to try to restrict production (Burawoy 1979; Edwards 1979; Montgomery 1979; Tulin 1984).

If manufacturing is in decline, is mass production? To address this issue, consider work as it is performed in different "megasectors"—broad groupings of different types of economic activity. Each of these sectors makes a distinct contribution to the economy—in extracting resources, in processing resources, in delivering goods, and in providing services.[2] The trends for employment in these sectors are represented in Exhibit 2.2, which shows the growing importance of service sector work, as well as the proportions of the labor force employed in other industries that seem (on the surface) to have helped society progress beyond mass production. However, consideration of many of the jobs within each of these sectors highlights how the typical strategies for organizing work in the manufacturing-based economy have been exported to other sectors and shape how work is performed outside manufacturing.

For most of human existence, most workers engaged in the extraction of raw materials—working in the areas of farming, fishing, forestry, or mining. But by the early 20th century, these workers composed only a relatively small segment of the workforce. Those few who remain on farms today perform work that bears little resemblance to the pastoral ideals of the family farm. Rather, most farming occurs as part of agribusiness, in which the methods of mass production have been applied to the raising of livestock, poultry, and produce (Schlosser 2005). The extension of mass production into farm work has required some farmers to learn to use advanced technologies to manage production. However, it also has contributed to the creation of a divided opportunity structure that limits prospects for workers (such as migrant farm laborers) to grow and advance. Mining remains an intensely physical activity, but one characterized by many

Exhibit 2.2   Employment Trends in Megasectors: United States 1940–2010

Source: Bureau of Labor Statistics.

"mass extraction" characteristics. Underground mining uses single-purpose machinery (such as "continuous miners") to grind the earth and place ore on conveyors. These technologies effectively work as assembly lines that run in reverse. Other forms of mining, such as strip mining (or even mountain top removal, in which whole mountains are demolished and the ore removed in the process) rely on heavy equipment and systematic processing of materials and operate on a scale that could only be characterized as "mass" production. All newer mining technologies require the application of greater skills than the older manual operations, but they also substitute massive equipment and sophisticated technology for human labor. This is one enduring persistent feature of the old economy, the drive to improve productivity by replacing labor through the use of complex technologies and managerial innovations.

The processing megasector focuses on the refinement of raw materials into finished goods, the intent of manufacturing and construction enterprises. As Exhibit 2.2 shows, processing work has declined significantly in the United States, which seems to support the "end of mass production" thesis. Yet, if we look at work within these sectors, mass production techniques have not been eliminated. It is true that there have been significant changes in this sector. For example, the traditional giant steel mill of the past has given way to smaller, more flexible "mini-mills" that produce smaller runs of more specialized steel products. But contemporary factories are still mass-producing consumer goods. And in certain parts of this sector, most notably the building trades, modern practices have made production more reliant on the use of standardized materials. Many homes are "prefabricated" in construction factories and simply assembled on site. Even the construction of "custom" homes (the largest of which are pejoratively termed "McMansions") depends heavily on cookie-cutter approaches to design and assembly. In short, flexible and custom production processes have taken hold in some areas, but they are being introduced in concert with a continued reliance on older systems of production, not simply replacing them (Pietrykowski 1999). Even the delivery megasector (which includes transportation, wholesale trade, retail trade, and utilities) relies heavily on mass production techniques. For example, the United Parcel Service (UPS) and Walmart (the world's largest private employer) use vast conveyor belt systems to sort and funnel parcels for delivery. Their successes are not built on unique products or customized services; rather, they are based on the application of mass production distribution techniques built with high technology and advanced accounting systems. Package sorting and shelf stocking are not skilled or challenging tasks that lead to career advancement; they are the type of routine work that one pursues as a means to earn a paycheck (McPhee 2005).

Finally, within the services megasector, which includes a wide array of enterprises—including leisure services, restaurants, hotels, real estate, financial, and public administration—mass production methods also can be found, often in highly developed, innovative forms. This sector includes some enterprises that do not rely on mass production—there are many relatively small enterprises in this part of the economy. Businesses such as high-end restaurants rely on workers' skills to produce a "unique" product for the consumer. However, this sector also contains many highly standardized operations that use the techniques of mass production to good effect. Each McDonald's, for instance, is little more than a small factory, composed of deskilled jobs designed to execute production of a limited array of standardized goods. Every aspect of the process, including the dispensing of condiments, the design of the stores, and the way in which customers are greeted, has been standardized so that the experience of eating (or working) in a McDonald's would be essentially the same wherever it was located. Even housecleaning teams are organized to work according to Taylorized methods, and the goals remain the same—to extract the maximum effort from each individual and minimize their chances of relaxing on the job (Ehrenreich 2001; Ritzer 2011).

In sum, there is a tendency to assume that the declining importance of manufacturing in the U.S. economy means that mass production is on the wane. It is not. Indeed, what has happened is that mass production techniques have been widely integrated into other sectors of the economy, including the rapidly growing service sector. Throughout the new economy, reliance on advanced technologies to perform complex tasks is expanding.

In some cases, these technologies have required an expansion in certain worker skills and transformed the ways in which work is performed. However, those same machines also have replaced many workers, both in manufacturing and elsewhere, and the persistence and spread of mass production methods cannot be ignored. As discussed below, the reality is that the new economy relies on both high-skill and low-skill work, and it uses both mass production techniques and newer, more flexibly specialized systems to produce goods and services.

New Skills?

The fact that the new economy is characterized by a new mix of industries and jobs has led some to conclude that it requires a new set of skills from the workforce. According to this argument, in place of the traditional low-skill work characteristic of the industrial economy, the new economy demands a workforce with new "soft skills" needed for successful social interaction in service sector jobs. It also demands a more highly educated workforce, as the emergent "knowledge economy" relies heavily on highly trained workers with substantial quantities of technical and intellectual skills. Some argue that, as these new skills have supplanted manual skills, the trend toward work simplification also has been altered—even reversed. This is because the predictable, carefully designed, strictly managed, routine jobs typical of mass production might be difficult to transfer to activities that require face-to-face interaction or sophisticated knowledge and problem-solving skills. Thus, the new workforce will have to be more highly educated and able to work independently, outside the strict controls imposed on routine manufacturing work. Let us examine some of these "new" skills to determine whether the claims being made about them can be supported.

Interactive Service Work

Interactive service work involves direct personal encounters between employees and customers. This requires the workers who perform these jobs to interact effectively with clients, and more often than not, to manage encounters that leave clients satisfied not only with the services rendered, but also with the experience of being served. While an increasing proportion of jobs require "people skills," it should be obvious that interactive skills are not specific to the new economy, nor were they completely absent from the old economy. It is true that many industrial jobs required few interaction skills, but there were other jobs (such as those involved in selling manufactured goods, managing manufacturing workers, doing the office work required by manufacturing enterprises, and so on) that clearly did. What this illustrates is that there is a range of jobs *within* industrial sectors. Not everyone employed in the manufacturing industry was a manufacturing worker, nor is everyone employed in a service industry performing interactive service work. Beyond this observation, it also bears noting that what constitutes service work varies widely. For example, the bedside manner that a doctor or nurse presents to patients is a critically important component of the work that they perform. In contrast, the depth of the interactive experience between a fast food worker and a customer is comparatively shallow, as are the expectations for emotional gratification.

One clear conclusion of a growing body of literature is that skills such as the abilities to communicate effectively with others, to present oneself appropriately and effectively, and to function effectively in a range of social situations are essential components of contemporary jobs and are valued highly by employers. For some workers, these skills have been cultivated through years of informal socialization, such as that received from parents who already know how to "fit in" with the types of people to be served. However, individuals from disadvantaged backgrounds often have developed very different types of interaction styles that lead to a poorer fit between their cultural tool kits and jobs. For example, a young man like Jamal from an inner city neighborhood may have a very difficult time "reading" the signals given off by the responses of a more privileged client, and he may well feel uncomfortable in the encounter. And the client may not necessarily treat the young man in the same way as someone perceived as being more similar in habits and disposition. These types of interpersonal dynamics present new challenges in fitting workers into the new opportunity structures, as the work is not only about mastering technical skills, but also social skills.

In theory, at least, it would seem that interactive skills would be resistant to routinization and simplification. However, as in the old economy, employers have attempted just that, and with considerable degrees of success. Perhaps the most familiar example of routinized interaction can be seen in the fast food industry. As Robin Leidner documented in her pioneering study of interactive service work, McDonald's has worked out very specific guidelines not just for the production of the food their stores sell to customers but also for structuring the interaction between workers and customers. Frontline service workers in the stores are given very specific scripts to follow and are instructed (and their performance is monitored) regarding what to say and how to say it. Nor is the management of interaction confined to employees, as McDonald's has succeeded in finding ways to encourage customers to behave in predictable ways; for example, by physically arranging the store to encourage customers to behave in prescribed ways (line up here, pick up your food there, don't linger too long, dispose of your trash on the way out), as well as limiting or scripting menu choices ("I will have value meal number 7"). As has frequently been noted, the experience of patronizing a McDonald's restaurant is more or less the same, no matter where it is located, as a result of this careful routinization (Leidner 1993).

Like routinization in manufacturing, this can be experienced as oppressive and intrusive by both workers and customers. There has been much speculation about the possibility that following scripts, and interacting with others according to externally imposed routines, rather than one's *own* reaction to the situation, can breed feelings of inauthenticity and psychological distress (Wharton 1999). But, it is also the case that efforts to make interaction more routine and predictable can actually help employees do their jobs, with the result that they sometimes embrace the protocols and actively collaborate with employers in imposing routines. Hochschild's (1983) classic study of flight attendants, for example, describes the various routines and scripts airline employees are trained to use during commercial flights. She points out that some of the reason for this training is to promote the corporate brand and to ensure that employees project the company's desired image. However, it is also the case that some of the routines help flight attendants perform difficult tasks: creating a reassuring environment that calms potentially panicky passengers, coping with angry customers when in flight (when there are limited options for discipline), and so on. Far from resisting or resenting routines, flight attendants adopt them willingly and make use of them voluntarily.

Thus, like work in the old economy, interactive service work is exposed to pressures to routinize and some tasks have been simplified and made predictable in very familiar ways. Once work is routinized, there exists the potential to replace people with machines. Can this be done within the service sector? Clearly this is possible. Consider, for example that inserting a credit card into the pump at a self-serve gas station is now the preferred means of payment, an act that completely eliminates the need to interact with the service station attendant. And it is possible to apply technologies to more sophisticated types of service encounters. However, there also appear to be limits. For example, one study of an attempt to implement an automated help desk found that computers have the capacity to respond to commonly experienced problems or needs of clients. However, what did prove to be an obstacle was that callers could not be controlled in the same ways that they are at a McDonald's restaurant. For example, callers to help desks typically did not know how to describe their problem in ways that the help desk computer could understand. Lacking technical sophistication themselves, callers often provided vague, incomplete, or even inaccurate descriptions of the problem they were experiencing, with the result that the computer was unable to ask many follow-up questions and lacked the capacity to respond appropriately (Head 2003). So, while employers continue to attempt to apply old economy techniques to the organization of work tasks, some of the complex services needed in the new economy may prove resistant to those efforts.

High-Tech Work

The new economy is also frequently described as requiring a more educated workforce. According to this view the "knowledge economy" or the "information economy" depends on a large and growing supply of highly educated, technically sophisticated employees, rather than the low-skill workers needed by mass production industry. An important element of this high-skill workforce is the so-called **STEM** (science, technology, engineering, mathematics) workforce; policy makers argue that economic growth increasingly depends on an adequate supply of workers trained in these areas, individuals who can design and maintain new technologies.

STEM jobs have, in fact, grown significantly. These trends need to be interpreted carefully, however—not all STEM occupations have been growing rapidly, or at all. STEM workforce growth slowed early in the 21st century in the aftermath of the dot-com bubble, and growth in the most recent period, while picking up, has been less rapid than the expansion observed in the late 20th century. Still, the growth has been real and projections suggest that the STEM workforce will continue to expand for the foreseeable future (Aspray 2010; Terrell 2007).

As the STEM workforce has grown, so have hopes that the new economy will create large numbers of good-paying, secure, challenging jobs. Enthusiastic visions of high-tech workers deeply engaged with creative work, making pioneering breakthroughs, and transforming knowledge into lucrative business opportunities are fueled by the success of enterprises such as Apple, Microsoft, and, more recently, Facebook and other social media. The demand for engineers, research scientists, and computer professionals, and the reality of success stories, indicate that these hopes are not irrational. However, there is reason not to jump to the conclusion that the low-skill mass production economy has been fully replaced by an economy dominated by autonomous, creative STEM workers.

First, it should be remembered that not *all* jobs in high technology, science-based sectors are well-paid, high-skill jobs. The computer industry employs many professionals, but also needs relatively unskilled workers to assemble, package, and transport its products. Similarly, while health care requires doctors, researchers, and other skilled professionals, no hospital can function without a significant number of orderlies, maintenance workers, aides, technicians, and clerks. Nor are the trends for the *professional* STEM workforce uniformly positive. For example, STEM salary growth has not outpaced salary growth in other occupations, suggesting that the demand for STEM professionals is not as great as had been anticipated (Hira 2010b).

In addition, the independent, creative, entrepreneurial STEM career exemplified by Steve Jobs of Apple and Mark Zuckerberg of Facebook is not the reality experienced by many employees in this sector. While earlier predictions (e.g., Kraft 1984) that high-technology work would be simplified and routinized in much the same way as manufacturing have proven exaggerated, STEM professionals work in jobs that are affected by a number of dynamics reminiscent of the old economy. For example, STEM employers in industries such as information technology (IT) can reduce the upward pressure on salaries produced by the shortage of well-trained workers by importing qualified employees from overseas. Much like manufacturing employers in the past, who encouraged and even sponsored immigration to keep labor costs down, contemporary IT employers look to countries such as India for sources of affordable labor. Moreover, the new economy offers novel opportunities to take advantage of the availability of lower-cost labor overseas. Many analysts have noted how the electronics revolution has made it possible to digitize steps in the work process and locate them offshore, retaining only portions of the work to be performed in the United States (Aspray 2010; Blinder 2006).

The actual work performed by STEM professionals generally requires considerable skill and expertise. However, it is not uniformly creative or autonomous in the way suggested by the more euphoric visions of the contemporary, high-technology workplace. Not all computer programming work, for example, is fully creative. Some computer professionals are engaged in the design of new systems and products and apply genuine creative autonomy as they enjoy the freedom to play with new ideas and concepts in their work. However, much computer programming (while it can be technically demanding) also can be relatively routine, formulaic work. For example, composing the well-understood computer code that forms the core of even new programs or checking and debugging programs can be (and often is) performed by lower-paid, and sometimes less-qualified employees. This is the type of work that is most likely to be outsourced to countries such as India (Aspray 2010; Parthasarathy 2004).

Finally, even the more creative kinds of programming work are often managed in ways familiar to the old economy. Rather than being given a free hand to design and create whatever and however they please, computer professionals typically have to design programs on schedules set for them by managers or external customers (in the case of suppliers and subcontractors to larger firms). They also often are required to incorporate user suggestions into their designs and their compliance with these and other expectations is closely monitored (Barrett 2004). The jobs of software developers are a long way from being the routinized work of mass production workers—they enjoy much more autonomy and have more creative decision-making ability than does anyone working on an assembly line. At the same time, as in the old economy, employers tend to seek ways to limit the cost of labor and attempt to manage and control it.

New Cultures of Control?

Some observers of the new economy have argued that changes in technology, organization, and markets are transforming jobs, including manufacturing jobs, into more highly skilled jobs in small, high-tech, "flexibly specialized" enterprises (Piore and Sabel 1984). The workforce needed for this new kind of workplace will be highly educated, multitalented, and required to exercise creativity and decision making on the job. In this vision of the new economy, managers are cautioned against "micromanaging" and are advised to form workers into teams with their own team leaders. Even low-level employees are hired as "associates," implying that their input will be valued. Others are not convinced. They point to the experiences of low-level service sector workers and agree with the critical philosopher Slavoj Zizek, who once quipped, "The employee of the month shows that one can be a winner and a loser at the same time." And, though acknowledging that new methods of organizing work are more common, these skeptics suggest that the reorganization of work is not just about introducing new markets or technologies, but also about developing sophisticated new methods to control workers and undermine their power in the workplace (Curry 1993; Parker 1985; Parker and Slaughter 1988). Thus, we must also ask, are new technologies and organizational designs increasing worker autonomy, creativity, and control?

Some of the most carefully conducted studies of this question have been performed by Steven Vallas, who has examined the introduction of computer technology and worker teams in the pulp and paper industry. His findings are important because they both support and refute core predictions regarding what happens to workers in organizations that use new technologies and managerial strategies. First, consider teamwork. Some analysts see worker teams as an ideological trick, a way of getting workers to believe that they have control when they do not. Vallas found that, rather than hoodwinking workers, work teams shared grievances, developed a heightened distrust of management, and developed class solidarities. And, by virtue of being a team, they expressed complaints to management with less fear of personal reprisal (Vallas 2003a). However, Vallas also found that the expectation for shared decision-making responsibilities between workers and managers has been exaggerated. In fact, the introduction of computer-regulating systems in the paper industry tended to increase the distinction between those who had the authority to make decisions and those who did not. Older workers interpreted the new technology as an affront to the craft skills they had developed through years of experience on the job, and they expressed dismay at the new reliance on meters and printouts that provided information that they already possessed (Vallas and Beck 1996).

One of the important insights from Vallas's studies is that the new economy seems to be marked less by fundamental shifts in the amount of control workers have, and more by modifications to systems of account-ability. In the old economy, the assumption was that workers should have no control and that jobs and machines should be rigidly designed and managed from above. In the new economy, where commitments to Total Quality Management foster a drive to work with exacting perfection, and where the responsibility for creating this outcome is placed on the shoul-ders of worker teams, a new dynamic of collective pressure is introduced. Workers now have increased responsibility for production, but not control over many of the decisions that shape it. These findings correspond with a number of other analyses that show that workplaces in the new economy do not operate on trust and cooperation, and that workers and managers remain skeptical of each other's intentions and motivations (Appelbaum and Batt 1992; Osterman 2001; Parker 1985; Rinehart, Huxley, and Robertson 1997; Smith 1990).

Another supposedly new dynamic is the re-creation of craft communi-ties, which some believe are reshaping the ways work is understood and performed. For example, the success of Silicon Valley enterprises has been attributed to their operating in a regional community of similar companies that specialize in computer work (Pietrykowski 1999). In some ways, these regional centers operate in a manner similar to the craft communities of bygone eras and position workers to exert control through the develop-ment of a collective culture of standards and expectations. The work of skilled programmers in this environment is truly a team effort and requires considerable coordination of tasks that require varying levels of skill and control. In this environment, workers are often paid quite handsomely for their efforts. On the surface, it looks like a model example of optimistic predictions about what work can become.

Questions have been raised, however, about how real these "craft communities" actually are. Even though many firms operating in Silicon Valley are small and nimble, they do not operate independently of larger corporate concerns (Harrison 1997). As we discuss further later in this book, some elements of life in Silicon Valley appear more enslaving than liberating. For example, programmers in Silicon Valley believe that it is their moral responsibility to labor 12- to 14-hour days and sacrifice their lives for their jobs. They work long hours not simply because of managerial demands or because of coworker pressure, but because they have internalized a sense of commitment to the organizational culture (Shih 2004). Even when they can take time out—for example, to care for a newborn child—they often do not because it would undermine their ability to be a "player" in their work community (Hochschild 1997). It is paradoxical that a work environment that offers control also tightens the chains of work, resulting in workers laboring far longer hours than those preferred by most individuals.

A final aspect of change to the cultures of control in the new economy concerns the application of **flexible work arrangements**. In the old economy, workers were commonly expected to be physically present at their work stations and to structure their time commitments in accordance with rigid daytime/full-time schedules. But as the labor force shifted to include increasing proportions of women, and as new technologies and organizational designs developed, many argued that jobs need to become more flexible. In stark contrast to the practices and philosophies central to the old economy, the thesis underpinning the implementation of flexible work arrangements is that the rigid schedules and physical copresence among workers are not necessarily the best ways to design jobs.

Before discussing the extent of flexibility present in jobs in the new economy, it is important to note that flexibility is not always intended to help ease the balancing of work and family or to maximize the liberating potential of technology. For example, the fast food worker who is sent home early because business is slack, or who is told to stay late because of an unexpected dinner rush, has a flexible work arrangement that operates primarily for the employer's interests. Sometimes organizational flexibility is used to legitimate dismissal of workers or justify the dismantling of work sites (Pollert 1988). These are not the types of flexible work arrangements that we are concerned with here. And as we discuss later in this book (especially in Chapter 4), this dark side of flexibility is potentially increasing.

The brighter side of flexibility is how it can be used to reconfigure job designs to suit the needs of workers. For example, sometimes workers seek flexible work arrangements to help them take care of a child or an aging parent. These types of needs might be satisfied by telecommuting, by reducing the number of hours worked, by scheduling work differently, or possibly by allowing a sabbatical from the job. Additionally, these arrangements may fit changing cultural preferences. For example, older workers commonly wish to remain integrated in the labor force, but to work fewer hours or to work part of the year. It is now clear that when they are applied with the employee's interests in mind, flexible work arrangements tend to help workers harmonize their lives with their jobs, as well as mitigate negative personal and family outcomes that result from work-family conflict (Kelly et al. 2008).

There are essentially three overarching types of flexible work arrangements. **Move work arrangements** include programs and practices that enable schedule variability, changes in starting and quitting times on a time-to-time basis, compression of workweeks, opportunities to work at home or off site, or to work at different worksites. Note that for these arrangements the amount of work remains the same, but the place or time at which the work is to be performed varies. In contrast, **reduce work arrangements** include programs and practices that enable workers to scale back work hours, to job share, to phase into retirement with reduced hours, or to work part of the year for a reduced amount of time. These options offer the prospect of temporarily limiting the amount of work to be performed. A third set of options can be considered **pause work arrangements**. These include programs and practices that enable workers to take temporary career breaks, sabbaticals, or paid or unpaid time for education or training to improve job skills. If employment opportunities are shifting in the new economy, we would anticipate that many of these move, reduce, and pause work opportunities would be available to workers. Are they?

To answer this question we draw on data from the 2009 Talent Management Study, a survey of 696 American companies, in which a key human resource officer reported a company's implementation of flexible work arrangements.[3] In respect to the various types of workplace flexibility, Exhibit 2.3 identifies the percentages of companies that offer the opportunity to move work, reduce work, or pause work to more than half of their employees. In many instances, companies that offered one type of flexible work arrangement also offered one or more other types of arrangements. With this observation in mind, we also identified the percentage of companies that did not offer any flexible work options to the majority of their workforces. As Exhibit 2.3 shows, over half of the employers (54%) provided no flexible work arrangements to most or all employees. It is also obvious that the availability of flexible work varies widely across sectors. Organizations in the manufacturing sector are the least likely to enable employees to move, reduce, or pause work, with 75% of organizations not offering any of these options to the majority of their workforce. In contrast, employers in the professional, scientific, and technical services sector and in the accommodation and food services sector are much more likely to offer access to flexible work arrangements (most commonly the option to move work in accordance with schedule flexibility). Among employers that offer flexible work arrangements, the most common approach is to enable workers to alter where or when they did their work (39%). Few employers provided most of their workers with the option to reduce work (15.8%) or pause work (16.1%).

The availability of flexible work arrangements also varies by the type of job being performed and by the type of worker performing that job. Ironically, the workers who often need flexibility the most—especially women—are the least likely to receive these options (Golden 2001; Swanberg, Pitt-Catsouphes, and Drescher-Burke 2005). Office administrative assistants, for example, are usually expected to work standard 9 to 5 shifts and to be at their desks during the workday, every day, even if work

Exhibit 2.3   Percentages of Companies Offering Flexible Work Options to Most or All of Their Employees

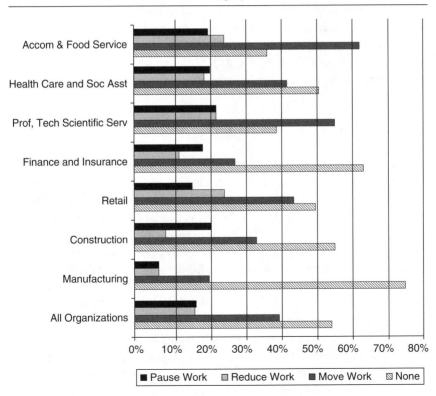

Source: Authors' analysis of the 2009 Talent Management Study.
Note: N = 696.

is slack. The managers for whom they work, however, may have numerous options to configure their schedules to meet their personal needs, to work at home, and some can even negotiate reduced work if they wish. The unevenness in the allocation of these types of opportunities is a critical concern in reconciling work and family for workers in all segments of the economy.

In sum, the existing evidence shows that some types of flexibility are more prevalent than others, but the option to reduce work or to take a break from work is not commonly provided to workers. Why is the availability of flexible arrangements so limited? Some evidence suggests that these practices help bolster workplace performance, facilitate retention, and enhance the job, and may even contribute to a company's reputation and shareholder value (Arthur and Cook 2004; Richman et al. 2008). And if this is the case for all employers, then the limited implementation may simply reflect a structural or cultural lag that likely will be reconciled eventually. However, at present there is conflicting evidence about the extent to which different types of flexible work arrangements consistently result in positive "returns on investments." (Kelly et al. 2008). In the end, employers are the ones who decide whether to extend flexibility to employees, and they look for evidence that the benefits of doing so (increased productivity, decreased turnover) outweigh the costs. In the absence of clear evidence of this result, it should be expected that the availability of flexibility will remain limited and uneven across industry sectors and across occupations.

The End of Organized Labor?

Another possible indicator of an emerging new economy is shifting balances of power between workers and their employers. One means of assessing this dynamic is to consider trends in collective action and unionization. As Exhibit 2.4 shows, the rise of the old economy was accompanied by a dramatic increase in union membership in the United States. At the middle of the 20th century, roughly one in three American workers belonged to a union. In the latter part of the century, however, membership plummeted, and, despite a short-lived increase in union membership in 2007 and 2008, by 2010, only 12% of workers belonged to unions. Overall, the decline of unionization in the private sector has been particularly sharp; just over 7% of private sector workers belonged to unions in 2009. The one exception to the general pattern of decline has been in the public sector. According to the Bureau of Labor Statistics, government sector unions remain vigorous and have even grown, representing more than 37% of government employees in 2009. But, as we discuss next, even in this sector the union movement is being challenged.

Exhibit 2.4   Percentages of American Workers Who Were Union Members:
United States 1930–2010

Source: Bureau of Labor Statistics.

One reason for falling union membership is that a declining proportion of the workforce is employed in manufacturing, the sector in which unionization has traditionally been strongest. In addition, as American manufacturing employment has contracted, employers have become increasingly willing to close facilities or to relocate both within the United States and abroad. In some cases, this allows employers to move away from areas where unions exist to those where they do not. In other cases, efforts to unionize have been stymied by threats to close shop and by actual relocation. However, nonmanufacturing workers do join unions in many other countries (including Canada). And the decline of unionization in the United States has occurred not just in sectors exposed to capital mobility, but also in sectors such as transportation and construction that cannot relocate. So, the real reason for declining unionization may be the failure of American unions to organize workers in the growth sectors of the new economy (Milkman 2006).

Much of the problem unions face in the new economy results from their limited success in attracting new members (Lichtenstein 2002; Moody 1997). American unions have been criticized for their lack of emphasis on organizing new groups in the post–World War II era, but even when they try, unions often meet with failure. One hindrance they face is that laws in the United States have made it comparatively difficult. Although American workers won the right to organize with the New Deal–era Wagner Act (which also established the basic legal framework for collective bargaining in the United States), numerous restrictions have since been placed on unions and union organizing activities. The Taft-Hartley Act of 1947 was particularly important in this regard. This piece of legislation, strongly backed by an antilabor, postwar U.S. Congress, prohibited some of the most effective strategies that had been used to expand the reach and power of unions, most notably the closed shop (in which all workers at a particular place of employment are required to be union members), the sympathy strike (in which workers in one industry strike in support of workers in another), and the secondary boycott (in which unions attempt to persuade others not to do business with a particular firm whose workers are on strike). Even more importantly, Taft-Hartley made organizing new groups of workers much more difficult by authorizing states to pass "right-to-work" laws (which prohibit unions from making paying dues or fees a condition of employment), by strengthening employers' ability publicly to speak out against and resist unionization drives, by permitting strikebreakers to vote in union certification elections, and by making the process of union certification far less flexible than it had been.[4] In general, U.S. labor policy has not favored new union formation, which in turn has left many workers (especially those in the rapidly growing service sector) without the opportunity to bargain collectively (Fantasia and Voss 2004). Moreover, employers have become increasingly hostile to unions and have successfully used a variety of tactics to discourage organization. In some cases, this involves the use of the "carrot"; that is, making union membership less attractive by providing, voluntarily, some or all of the good labor conditions that unions gain for their members (Milkman 2006). In others, it involves the "stick"; that is, aggressive efforts to block unionization through methods such as firing organizers and the use of experts on keeping unions out (Goldfield 1987; Head 2004).

Government workers in the United States have not had to contend with strong antiunion efforts by their employer. Though some states limit public sector unions in various ways (e.g., New York State's "Taylor Law," which prohibits strikes by public employees), government workers in general have encountered fewer obstacles to unionization than their private sector counterparts, which probably accounts for much higher rates of unionization in the public sector. However, in the aftermath of the financial crisis of 2009, a number of state governments, including states with strong union traditions such as Ohio and Wisconsin, took aggressive action to curtail public sector unions. Arguing that the salaries and benefits paid to current and retired government workers were crippling state budgets, and complaining that union job protections limited flexibility and prevented institutions such as public schools from rewarding merit, governors such as Scott Walker of Wisconsin and John Kasich of Ohio led movements to impose legislative restrictions on public sector unions. Calls for similar limits on public unions have been heard in some European countries. Time will tell if these movements will achieve their objectives. Survey data indicate significant support for public sector unionization rights, and efforts to overturn anti–public sector union legislation are under way. This indicates that the path to limiting the power of collective bargaining in the public sector may be politically risky. However, if state governments succeed in limiting public sector unionization, that almost certainly will negatively affect the terms of employment for public sector work (Greenhouse 2011a; "Public Sector Unions" 2011).

Some analysts have suggested that the new economy has the potential to change relationships between employers and employees for the better, and that these new employer-employee relationships reduce the need for the checks that unions place on employers' inclinations toward exploitation. For example, new management philosophies and new forms of work organization ostensibly promote collaboration and harmony between employers and their workers. Even if this is not evidenced in reality, the belief that this is the case can have significant power. Work in smaller enterprises involves direct face-to-face relationships between workers and their employers, which discourages a "them and us" mentality, and as greater shares of workers are laboring in smaller enterprises, this might be having an influence. In this culture of cooperation, traditional unions are not attractive to workers, who often see them as threatening the company's interests and their jobs (Heckscher 1988). Workers also are less likely to devote their entire careers to single employers, and instead they build portfolios and move from employer to employer. In this context, unions representing particular workplaces become less relevant (Cappelli 1997). But this is only part of the story.

The decline of unionization has occurred simultaneously with declining wages and job security. Thus, it may not be that changing management practices are making unions unnecessary; rather, it may be that unions have lost much of their ability to negotiate for workers' interests or to persuade workers that they are better off with unions than without them. Some evidence suggests that unions have lost much of their teeth. For example, American unions' ability to use the strike to put pressure on employers to raise wages or improve work conditions has largely evaporated. Consider that, according to the Bureau of Labor Statistics, in 1970, 2,468,000 workers participated in 381 mass walkouts. In contrast, in 2010, there were only 11 large-scale work stoppages, and these only involved 47,000 workers.

The downward trend in union membership has been witnessed as well in most countries outside the United States, but the extent of these declines has been variable. Scandinavian countries have retained remarkably high union membership. Countries such as Australia, Germany, and the United Kingdom have significantly higher union membership rates than does the United States. Neighboring Canada has unionization rates more than double those of the United States (Exhibit 2.5). The United States stands out in the international arena because of the combination of dramatic declines in membership and its exceptionally low current levels of unionization. These declines have contributed to overall declines in wages, benefits, and job security for workers (Lichtenstein 2002). The fact that unions remain important in other advanced economies, and that service workers who are typically unorganized here belong to unions there, is strong evidence that unions can still matter. Unions are not just a vestige of the old economy, but whether they will thrive in the new economy depends on whether they develop effective strategies for attracting new members and organizing new sectors of the workforce (Clawson 2003; Fantasia and Voss 2004; Heckscher 1988; Milkman and Voss 2004). It also depends on the resources made available to union organizers and the constraints placed on their actions.

Exhibit 2.5   Trade Union Members as Percentages of All Employees: International Comparisons

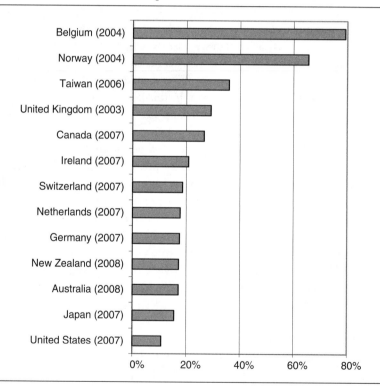

Source: Hayter, Susan and Valentina Stoeveska, 2011. Trade Union Density and Collective Bargaining Coverage: International Statistical Inquiry 2008-09. International Labor Organization. http://www.ilo.org/public/english/dialogue/ifpdial/index.htm.
Data provided to authors from the International Labor Organization.

A New Global Economy?

A final and much-discussed aspect of the new economy concerns globalization. Has the emergence of a global economy fundamentally changed the U.S. economy and the situation of workers worldwide? In some respects, the global economy is not really new. The histories of virtually all modern societies, from the 16th century onward, can be traced to international economic ties (Wallerstein 1979, 1983). For example, colonial America participated in international trade of slave labor, sugar, rum, tobacco, cod, and textiles (Kurlansky 1998). Had it not been for these exchanges, the present-day demographic makeup of the United States would be profoundly different, as would its culture (Breen 1985; Sobel 1987). Likewise, the export of slaves to the United States had an enduring impact on the development of African societies (Fredrickson 1981; Portes and Walton 1981). Trade with Asia is not new, and the European "discovery" of America was the result of attempts to find better trade routes. Indeed, international trade has long been in existence and has gone through numerous cycles of growth and decline (Chase-Dunn, Kawano, and Brewer 2000).

Nevertheless, one must still acknowledge that the extent of global economic activity is unprecedented and that the penetration of global capitalism to all corners of the world is both more complete and more complex than ever before. The new global economy can be described as a vast international network capable of rapidly developing and diffusing resources, technology, and information across the world. Among the key characteristics of the new global economy are the following:

- The immense volume of trade and consumption between societies;
- The rapid transmission of information between societies;
- Powerful and transportable technologies implemented throughout the world;
- Intense "dis-integrated" production, spread over national boundaries; and
- Flexible arrangements that enable employers to shift production and consumption from one society to another.

It should be added that the new global economy also is characterized by new types of legal and political arrangements designed to promote globalization. In the past, global economic activity was supported politically in various ways. Nineteenth-century colonial occupation created economic opportunities for employers in the dominant country (e.g., British enterprises benefitted greatly from British colonial activity in places such as India). Even after the colonial empires dissolved, enterprises in the developed world often continued to derive considerable benefit from activities in the former colonies, aided by political, military, and diplomatic support from their home governments. On the other hand, international economic activity, especially among industrialized countries, was often limited in various ways by tariffs and other legal arrangements designed to protect domestic economies. It is this last situation that has begun to show signs of change in the new economy. Accompanied by economic theories that trumpet the benefits of free trade and the free movement of goods, services, and investment, many countries have joined in a variety of free trade agreements and even currency unions that remove many of the barriers to international economic activity. Notable examples include the North American Free Trade Agreement, designed to enhance free trade among the United States, Canada, and Mexico, and the European Union, which has created a large free trade zone among many of the countries of Europe and includes a common currency shared by at least some of its members. These arrangements make it much easier for goods and services to move across borders and for companies based in one country to establish facilities (or move them) to a range of others. One result has been that cultural and technological changes in developed and developing countries are accelerating at unprecedented rates. The speed of change, the extent of diffusion, and the flexibility of webs of connection set the current organization of work apart from the systems that preceded it (Castells 2000; Mattsson 2003; Milberg 2004). As one observer has remarked, the world is now, probably for the first time, approaching "universal capitalism" (Wood 2003).

Companies are integrating themselves into the global economy for a variety of reasons. Labor cost savings are a major motivator, but other considerations are involved also. Developing countries almost universally have lax environmental regulations, and this reduces the expense entailed in limiting and controlling pollutants (Jones 2005). The tax structures of nation-state systems also encourage the movement of jobs. For instance, when the industrial tool company Ingersoll-Rand relocated its headquarters to Bermuda, it reduced an annual $40 million tax bill owed to the United States government to $27,653 (which it paid to its new home country) (Johnston 2004). Companies also internationalize operations to secure government contracts and extend the global reach of their product lines.

The emergence of a global economy has had significant effects on workers in the United States and other industrialized countries, partly because the "national" character of companies has weakened and become ambiguous. American companies such as Ford have long had overseas operations in Europe. However, the scale of those overseas operations is something quite new. Ford (like nearly all the major employers in the new economy) is a multinational corporation that operates both within and beyond the political realms of nation-states. Today, Ford employs more workers outside the United States than it does within, and it manufactures and sells cars around the world. Companies such as Ford can easily shift production from place to place, as workers in Michigan have discovered. When companies move or open new facilities abroad, they bring more than jobs—they also sometimes spread culture and methods of organizing work. A good example of this can be seen in the case of Japanese companies, which have built facilities in both the industrialized and developing worlds. These plants became vectors through which Japanese production methods diffused to places such as Marysville, Ohio, Northern Mexico, or Spain (Elger and Smith 1994). However, in other cases, foreign companies adopt local practices. A recent editorial in the *Los Angeles Times* argued that European companies, when they establish facilities in the United States, actually adopt American-style practices (such as "union-free" workplaces) that differ radically from the same companies' practices in their home countries (Meyerson 2011).

The emergence of a global economy has undoubtedly brought with it much that is new. Workers in the United States are much more likely to encounter technologies and managerial practices that originated elsewhere, and workers in developing countries have been drawn into more direct relationships with global webs of production. Still, arguing that all this is *entirely* new seems an exaggeration. Employers have relocated in the past, and even the earliest industrial firms in America "borrowed" practices from the pioneering British. It seems more accurate to say that globalization has accelerated and intensified existing dynamics at work. At the same time, the socioeconomic differences between the developed and developing world have not been erased by these changes (an observation that we document in the next chapter). Nor have national differences in workplace practices been eliminated by globalization (Smith and Meiksins 1995). As with the various other changes we have reviewed, there is much of the old within the new economy and workplace.

Conclusion

Does the existence of new products and new ways of working verify the existence of a new economy? The answer is both yes and no. The U.S. economy has shifted away from a primary reliance on its manufacturing base, and the expanding service sector is creating new job demands, as well as opportunities. So the economy is new in respect to the diversity of goods and services produced. But in many of the new jobs and industries, old ways of working and managing work are applied, indicating the persistence of the old economy. While some of the new jobs require different skill sets, substantial portions of the labor force work in low-skill jobs that require few skills, again indicating that not everything has changed. While organizations have redefined manager-worker relationships and are relinquishing more control to workers, some of these changes appear to be window dressing, and in other circumstances have not led to a true enhancement of job quality. While the new economy offers prospects for flexible work, availability is limited and uneven. And while the global economy has shifted work opportunities, some of these changes are not so much new as they are extensions of tendencies that have been present for centuries.

We argued throughout this chapter that, while there is a new economy, it has been laid on top of and has not replaced the old economy. We also argue that the same forces that underpinned the development of the old economy continue to shape many of the contours of the new economy. The new economy offers great opportunities to liberate work, through new applications of technology, through new organizational designs, and by harnessing the contributions of a diverse workforce. But it also contains many elements of the old economy that block these opportunities and perpetuate and reproduce old ways of working. Both optimistic predictions that we are entering a new world of work and pessimistic comments about "old wine in new bottles" miss the complexity of contemporary work. Our analysis indicates that there will be both change and continuity in the emerging new economy, and that there will be both positive gains for some workers and continued hardships for others.

Notes

1. The Bureau of Labor Statistics reports that in 2010, 3,283,950 people were employed in computer and mathematical occupations in the United States, and 11,027,340 in food preparation and serving related occupations.

2. Economists use the terms *primary, secondary, tertiary,* and *quaternary sectors.* However, because they also use the phrases *primary* and *secondary labor markets* to describe divided opportunity structures, we use the terms *extraction, processing, delivery,* and *service provision* in accordance with the divisions identified by Kenessey (1987).

3. For further information on the Talent Management Study (Marcie Pitt-Catsouphes and Stephen Sweet coprincipal investigators) see the industry reports published by the Sloan Center on Aging and Work at Boston College (www.bc.edu/research/agingandwork/projects/talentMgmt.html).

4. The United States employs a cumbersome, two-step process for union certification in which workers must first sign cards indicating their interest in forming a union, then, at a later date, vote. Under the Taft-Hartley Act, all of this must take place under the watchful bureaucratic eye of the National Labor Relations Board. At the time of this writing, a proposal to substitute a one-step union certification process had stalled, despite initial support from the Obama administration.

Chapter Three

Economic Inequality, Social Mobility, and the New Economy

One would hope that the transition to the new economy is helping more families afford a better quality of life. One would hope also that it is increasing the opportunity to obtain quality jobs, the kind of work that offers the opportunity for growth and advancement, as well as economic rewards. Better yet, one would hope that these improvements would be evident not only in the United States, but also elsewhere in the global economy. And, finally, one would hope for equity, such that no one segment of the population reaps an inordinate share of the benefits or bears most of the risks and burdens. To what extent is the new economy fulfilling these hopes?

Social scientists have long argued that changes in work opportunity are intimately related to patterns of inequality and opportunity, but in anticipating where societies are headed, theories often make very different predictions. Consider, for example, the debate about the effects on inequality that occurred in the early stages of industrialization. Karl Marx (Marx and Engels 1972) argued that industrialization brought with it high levels of inequality and class antagonisms, and that the ruling class (the bourgeoisie) obtained its wealth by appropriating it from the labor performed by the working classes (the proletariat). In predicting the future of the old economy, Marx believed that an oppositional class structure, the driving motor of capitalism, would polarize the structure of society into two classes, a small affluent class and a large exploited and disadvantaged class. In contrast, Frederick Winslow Taylor ([1911]1964) denied the existence of antagonistic class interests and argued that industrial capitalism would

benefit all. In defending scientific management and the practice of dismantling the work of skilled craftsmen into repetitive unskilled jobs, Taylor
argued that even if workers labored in low-skill jobs, they would be better
off economically, as their companies would be more productive and create
more wealth to distribute to everyone. Today, we are positioned much like
Marx and Taylor were in the early stages of industrialization, as they tried
to figure out the impact that a new economic order would have on the
future of work and opportunity.

How is the emergence of the new economy reshaping patterns of
inequality and opportunity? Are the decline of manufacturing and the
rise of services liberating workers from economic hardship and spread-
ing greater affluence to all? Or is it imposing new limits on opportunities and hardening already existing lines that separate the classes? Is the
growth of industry in the developing world producing rising standards of
living and more opportunities, or is it primarily benefitting a small elite
and leaving existing class inequalities intact? In this chapter, we examine
these questions, exploring how the changing contours of work affect
structures of opportunity and the conditions under which the world's
people live.

Are Economic Divides Narrowing or Widening in the United States?

Economic success is commonly gauged by a single measure, such as per
capita income or median household income. While useful, these broad summary measures fail to capture the diversity of outcomes that any event may
create, as it is possible for a society to have a very high *average* income,
but for many individuals also to be excluded from pursuing opportunity.
For this reason, throughout this chapter we disaggregate economic data so
that subgroups in the society are compared relative to one another, as well
as over time. Our first concern is to identify the scale of social inequality
as it currently exists in the United States and to gauge how that compares
to the inequalities evident in the old economy. Our measures of the new
economy rely on the most recent indicators available, and we mark 1970
as a benchmark point from which to evaluate the opportunities present in
the old economy. As discussed in the previous chapter, this year approxi-
mates the point at which the mass production industrial economy in the
United States started to shift toward a more diverse postindustrial economy.
With that shift, union membership declined, organizational practices were
reconfigured, new flexible technologies were introduced, and (with the

assistance of new technologies) the globalization of production took hold. This benchmarking year also presents the old economy in perhaps its most favorable moment, as job security and wages were high for many workers, wage growth was nearing its plateau, and unemployment was low. Many of those features changed in the decades that followed, especially for blue-collar industrial workers in the Northern rust belt, as factories closed during a period of deindustrialization in the 1970s and 1980s (Bluestone and Harrison 1982). Today, some of these communities have successfully made the transition to an employment base rooted in the new economy (e.g., Pittsburgh, Pennsylvania) while others suffer from stagnation or ongoing decline (e.g., Youngstown, Ohio, or Flint, Michigan).

In some ways, the situation present in 1970 affirmed the perspectives of both Taylor and Marx. In support of Taylor, the movement toward mass production work (even that which was deskilled) resulted in substantial wages and security for those employed in the manufacturing sector. Because the manufacturing sector was the bedrock of the old economy, the archetypal worker was someone like Dan (the automobile worker introduced in Chapter 1). But the followers of Marx can rightly claim that the economic returns that manufacturing workers received were not inevitable outcomes of productivity. They were as much a consequence of organized labor acting in its own class interests. And while manufacturing workers had high wages and security, many others (especially women and minorities in the lower classes) were funneled into jobs that offered far fewer opportunities. The question we consider here is whether the shift to the new economy has resulted in a fundamental shake-up of the amount of class inequality present in American society, and if so, which groups have benefitted most.

Exhibit 3.1 offers a graphic illustration of the changes in incomes in the United States, examining the earnings received by five equal-sized portions of households (termed *quintiles*). To make comparisons over time possible, all of the dollar figures have been adjusted to their purchasing value in 2009. The darkest shaded band at the bottom of the graph shows that, in 2009, the poorest fifth of households in the United States subsisted on incomes ranging from $0 to $20,453, the next fifth had incomes that ranged from $20,454 to $38,550, and so on. One-fifth of households in 2009 had incomes higher than $100,000 (the top white-shaded portion of the graph). Because the range of incomes for these top earning households extends into the millions of dollars (for people like Bill Gates), we capped the graph at a maximum income of $200,000, but added a line representing the threshold that separates the top 5% of households from everyone else ($180,001 in 2009).

Exhibit 3.1   Incomes Received by Each Fifth of Families: 1967–2009 (Prior Years Converted to 2009 Dollar Values)

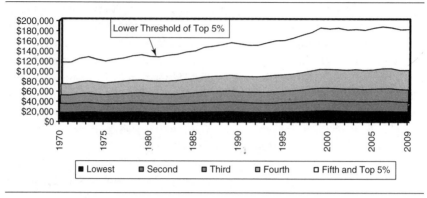

Source: U.S. Census Bureau (see www.census.gov/hhes/www/income/data/historical/inequality/index.html).

To gauge the impact that the transition to the new economy had on the relative earnings of different strata, consider the relative positioning of each fifth of the population in 2009, as well as how the maximum income threshold for each quintile changed in comparison to 1970. Exhibit 3.1 reveals that the threshold separating the poorest fifth of households from the rest of society increased by only $2,273, from $18,180 in 1970 to $20,453 in 2009. That means that roughly one in five households was living below or near the poverty level in 2009, and that this proportion has remained largely unchanged since 1970. The second quintile of households had maximum incomes of $38,550 in 2009. Again, we observe only a modest change in their overall economic condition, as the threshold that separates them from the next quintile only moved upward by $3,723 from 1970 to 2009. This stratum of households, like their counterparts in 1970, remained situated in a position that offers few possibilities to accrue savings and puts homeownership largely out of reach. Households in the third quintile had a maximum income of $50,656 in 1970, and that threshold rose by $11,145 to reach $61,801 in 2009. For the top two-fifths of households, however, evidence suggests substantial income growth. Households in the second highest quintile saw maximum incomes rise from $72,273 in 1970 to $100,000 in 2009. Clearly the greatest change in incomes is reflected in the top-earning households. For the richest one in 20 households, the lower threshold that separated their earnings from those received by everyone else rose from $114,243 in 1970 to $180,001 in 2009, a 58% growth rate.

One can present an optimistic case that the new economy is creating prosperity and argue that a rising tide raises all ships. However, a more careful reading of the data reveals that the really significant gains in income went to top earners. In contrast, the lowest two-fifths of all households continued to earn incomes similar to what households earned in the early 1970s. While there has been some modest income growth even for these strata of households, as well as for the third quintile, the conclusion that they are modestly better off is open to debate, given that the incomes achieved are sometimes the consequence of working harder either to maintain one's standard of living or to satisfy new needs critical to career and family success in the new economy. For example, one means of replacing declining wages is to have a spouse enter the labor force or for one or both partners to increase their work hours. According to Jerry Jacobs and Kathleen Gerson (2004), if one measures the total amount of labor performed by the typical American household, American families are now working the equivalent of an additional month each year as compared to the amount of work done in 1970. And to keep the income stream intact (most commonly by replacing a single earner household arrangement with a dual earner arrangement) requires making additional investments, such as purchasing a second car or allocating money for child care (Warren and Tyagi 2003). The key point to understand here is that even if the lower three-fifths of households are making modestly more than they did in 1970, they are not substantially better off in terms of their economic situation. They may be earning a little bit more, but they are working much harder to do so, and usually because they need to.

With the help of Exhibit 3.2, we consider further the issue of equity. These graphs indicate what proportion of total income went to different groups of households in 1970 and 2008. Looking first at 1970, observe that the richest 5% of households took home nearly one-fifth (17%) of the total income earned that year. If we add that figure to the share of income received by the remaining members of the highest quintile of the population (27%), we find that the top fifth of households received 44% of all of the income earned in 1970. Compare the proportion of earnings received by these top-earning groups to that received in 2008, when the richest quintile took home half (50%) of the collective income earned. The rich have gotten richer. In contrast, those in the bottom tiers of the class structure are receiving a smaller piece of the income "pie." The poorest fifth of households received only 3% of the total dollars earned in 2008, a smaller share than was received in 1970. Similar declining shares are observed for the second and third quintiles.

Perhaps there is no better case than Walmart (now the world's largest private sector employer) to illustrate the issue of compensation equity and what this means in the lives of individuals. Our friend Jodi has worked at Walmart for 3 years and is now considered to be an exemplary senior employee. Most commonly she can be found either at the register or behind the

Exhibit 3.2   Share of Aggregate Income Received by Each Fifth and Top 5% of
Households: 1970 and 2008

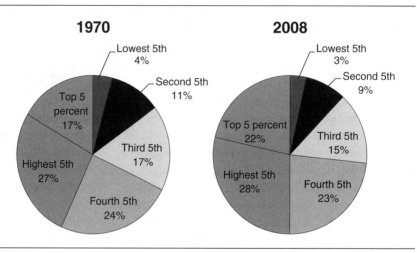

Source: Statistical Abstracts of the United States.

returns counter. Jodi can request days off, but those requests are not always accommodated, and she earns the same amount whether she works days, evenings, or on weekends. Her schedule changes week to week, as do her hours. Because she had solid evaluations, reflecting a positive attitude and diligence, her initial wage of $8.00 an hour increased by 40 to 60 cents per hour in each of her years of service. After three years of work for Walmart, she now earns $9.60 an hour, still well beneath a living wage, estimated to be slightly above $11 an hour in the local community. While she wants to work as many hours as possible, she is never scheduled to work full time because that would require her employer to provide additional benefits. If she arrives late or fails to come in as scheduled, she loses the few privileges that she is awarded on the basis of her seniority with the company. Many of Jodi's coworkers have significant health concerns that can be directly traced to financial hardship (e.g., missing or decaying teeth), but her most recent pressing concern is figuring out how to keep her car on the road.

Contrast Jody's meager income to that of Walmart's president and CEO, Michael T. Duke, whose total compensation was $19,234,268 in 2009 (according to AFL-CIO Executive Paywatch). If Jodi were to secure full-time work at Walmart (which she cannot), at her current wage rate she would have to work 1,308 years to earn what Michael earns in a single year. Conversely, in a single morning Michael earns far more than Jodi's annual income. Walmart is not exceptional in its differential treatment of employees at the high and low ends of its organizational structure. The

AFL-CIO reports that in 1980 the average CEO made 42 times the pay of the average employee, but in 2009, the ratio had increased to 263 to 1. Nor is this just an American phenomenon. While the gap between CEO and worker pay is much higher in the United States than in other industrialized countries, the gap is still large elsewhere. Even egalitarian Sweden had a CEO to worker pay ratio of 19 to 1 in 2005. Calls for limits on CEO pay increased during the recent economic downturn, but, so far, there is no compelling evidence that real changes have resulted as a consequence of attention to this issue (Mischel, Bernstein, and Shierholz 2009). These types of class divides, those separating the incredibly affluent from the rest of society, reflect the concentration of wealth and power that already existed in the old economy (Mills 1978). However, the chasms that separate those who have from those who do not are widening, as the data consistently indicate that the richest segment of the population has received an increasingly large share of total earnings since 1970.

The terms of Jodi's employment place a burden on society as a whole, not just on Jodi. While her low wages make it possible for consumers (both the poor and the affluent) to purchase items at a lower cost than would otherwise be possible, the inability of low-wage workers like her to afford health insurance not only affects their lives, it also results in higher medical costs for everyone, as these low-wage workers (and their children) tend to seek medical attention only when it reaches critical conditions. To afford housing, they need heat and rent assistance. The fact that they cannot afford child care increases reliance on remedial services and correctional services. The costs of these (and other) services are increasingly carried by members of the middle classes, not by those with the fastest-growing incomes (the wealthy) or by corporations. As reported by the Tax Policy Center of the Urban Institute and the Brookings Institute, in 1970, the marginal income tax rate on the highest earners was 71%, but by 2011, this had fallen to 35%.[1] While official income tax rates are structured so that the very affluent pay a larger share of taxes than the other income groups (owing to their larger incomes), the amount received is substantially less than that which could have been expected had their contributions remained at a level consistent with the standards applied in the old economy. And as Warren Buffett observed, because of various tax exemptions, the fabulously rich usually pay a *lower* share of their income than do members of the middle class. Corporations now contribute a smaller share as well, as their marginal effective tax rate declined from 42% in 1970 to 23% in 2003 (Tax Policy Center 2011). Beyond that, corporations have become remarkably adept at skirting tax obligations. In 2010, for example, General Electric earned a profit of $14.2 billion, and nearly one-third of those profits came from work performed in the United States. Yet, not only did General Electric not pay any taxes, it actually claimed a tax benefit of $3.2 billion (Kocieniewski 2011).

In sum, the transition to the new economy has economically benefitted those at the top of the class structure far more than those at the bottom. The extent to which the middle fifth of households have benefitted from the transition is debatable, but it is clear that these households now have additional expenses that have resulted from new demands, and they also are working harder to earn what they have. It is also clear that the lowest two-fifths of households are not significantly better off in comparison to where they stood in the old economy.

Are Career Pathways Opening or Closing?

The transition to the new economy has been accompanied by growing income inequality and more highly paid workers have done far better in recent decades than those at the bottom of the income scale. But, we must ask whether these trends are actually *the result* of the new economy. What is it about the new economy that has produced the higher levels of income inequality we are seeing? Or, has income inequality grown independently of changes in jobs? Patterns of income inequality varied at different periods of the old economy—perhaps what we are seeing is just another one of those variations.

Obviously, some of the changes in income inequality occurred because of factors such as government policy (e.g., not raising the minimum wage to keep up with the cost of living) and corporate policies (increased stockholder willingness to authorize huge compensation packages for CEOs) that are not directly related to the emergence of the new economy. However, there is good evidence that the characteristics of the new economy *are* responsible for a significant portion of the growth of inequality. Most important have been changes in the jobs that characterize the new economy and the skill demands associated with them.

As indicated in Chapter 2, one of the most important characteristics of the new economy is the shift away from manufacturing work toward service employment. The factory jobs that used to be plentiful in the American economy have become harder and harder to find, while employment growth has occurred primarily in a wide range of service occupations and industries. Economists and sociologists who have analyzed the service-dominated new economy have concluded that the mix of occupations it creates is the key to understanding changing patterns of income inequality. Specifically, they argue that the service economy has tended to produce a more polarized workforce—highly paid jobs have not disappeared, but getting them now requires higher levels of education and significant levels of skill. Well-paid jobs for those lacking advanced education have become less common, along with the factories where they used to be located. In their place, poorly educated workers now find only low-level jobs that offer little in the way of pay or opportunities to advance.

In short, the polarized distribution of income we have identified is directly related to a polarized structure of opportunity in the new, service-dominated economy (Autor, Katz, and Kearney 2006; Holzer et al. 2011; Moller and Rubin 2008; Mouw and Kalleberg 2010). Central to the new economy is the **dual labor market,** a structure that divides opportunities into two different strata of jobs. The favored jobs are located in the **primary labor market,** which offers higher pay, predictable career structures, opportunities for skill development, and enhanced job security. These are the good jobs. In contrast, other workers are trapped in a **secondary labor market** comprised of jobs that offer low pay, unstable opportunities, low security, and limited skill expansion (Piore 1977). Most of the low-skill jobs that workers hold today are nestled in the secondary labor market, offering neither significant material rewards nor stepping-stones to greater opportunity. Like Jodi's job at Walmart, these are dead-end positions that lack benefits such as health insurance, provide little security, and offer scant economic reward (Kalleberg, Reskin, and Hudson 2000). It is important to note that divided opportunity structures developed within the old economy, and also that aspects of the division between the primary and secondary labor markets appear to be changing—not because bad jobs are becoming good jobs, but rather through the extension of insecurity into some jobs in the primary labor market (making them more like secondary labor market jobs). As we discuss later in this book, some workers—even those in good jobs—are working longer hours, facing new forms of insecurity, and experiencing greater challenges in locating work than was the case a few decades ago. Nevertheless, the new economy is deeply divided and the availability of good jobs in the middle, particularly for those with limited education, is increasingly limited (Perrucci and Wysong 2002).

In the old economy, a dominant career path was to expect income growth and security with advancing age (Cappelli 1997). While the prospects of obtaining a good job were unevenly distributed on the basis of race and gender, it was not unusual for workers with low levels of education to land, if not interesting jobs, ones that would (over time) eventually enable them to afford a middle-class standard of living. According to the economist Frank Levy (1998), because the economy was expanding the earnings potentials for all social classes, the experience for workers in the decades that followed World War II was akin to walking up an upwardly moving escalator. As we discussed above, in the new economy, the escalator now seems to be moving along at a brisk pace for those at the top of the class structure, it might be moving at a slow pace for the upper middle class, but it is not moving at all for those in the middle and lower classes. The question we consider here is the likelihood of individuals' being able to jump from one escalator to the next, to move from jobs that offer few prospects for advancement into jobs that do, or to walk up whatever escalator they are on (even if it is stalled or moving backward). We draw attention to two types of mobility, intra-generational upward mobility (the prospect of moving up a career ladder within one's lifetime) and intergenerational upward mobility (the prospect of landing jobs better than those held by one's parents).

Missing Rungs in the Ladder

It is clear that many workers in the new economy cannot "work their way up" in the manner that was evident in the old economy, and the entry point is increasingly important in determining the range of jobs that one might eventually be eligible to hold. For most workers, economic security in the new economy hinges on the attainment of a college or graduate degree (Levy 1998; Marshall and Tucker 1992; Reich 2002). As a result, Americans are entering college programs and returning to school at unprecedented rates. Exhibit 3.3 shows that, in 1970, only one in two Americans earned high school degrees; today this figure has risen to nearly nine in 10. The proportion of adults who are college graduates rates has tripled during the same period, as younger Americans are increasingly likely to seek postsecondary education. In 2008, 40% of young adults (age 18–24) were enrolled in college (Chronicle of Higher Education 2010).

Exhibit 3.3   Educational Attainment of Adults, Age 25 Years and Older: United States 1970–2009

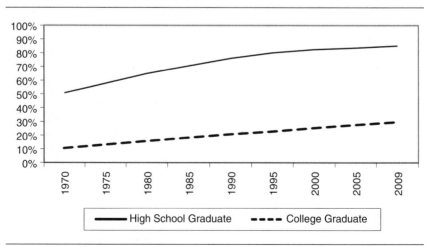

Source: Statistical Abstracts of the United States.

The dramatic increase in entry into college reflects personal responses to changing opportunity structures. And it is not only young adults who are investing in a college education in hopes of career success. Older Americans are returning to school in unprecedented numbers. More than one in three students (37%) enrolled in colleges and universities in 2008 was age 25 or older, and most of these students were women (Chronicle of Higher Education 2010). Among these nontraditional students, the most common reason for returning to school is to relaunch careers or adjust skills to keep up with today's job markets. While many older students find returning to school intellectually and socially invigorating, the combined burdens of student, worker, and parent roles place considerable strains on marriages and family lives (Hostetler, Sweet, and Moen 2007; Sweet and Moen 2007).

As Exhibit 3.4 shows, in 2008, the average worker possessing a bachelor's degree earned more than twice the income of a high school graduate, and those possessing graduate level degrees earned considerably more than those who ended their education at the bachelor's level. Those possessing only high school diplomas typically hovered at or near the poverty level, and the economic return on a limited education has declined during the past four decades. If the earnings standards for high school graduates in 1970

Exhibit 3.4   Educational Attainment and Average Worker Earnings: United States 2008

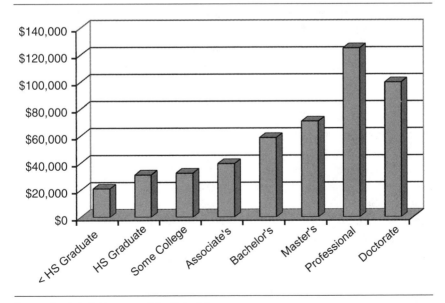

Source: Statistical Abstracts of the United States.

had remained constant, these lower educated workers would have been earning $44,000 in 2008 ($12,000 more per year than what was the actual average income earned).

Those who ignore the structural arrangements in the new economy argue that growing inequality is largely a result of choices—those who make the wise decision to attend college get good jobs, whereas those who do not fall behind (e.g., Herrnstein and Murray 1994). There are indications, however, that a college degree is no longer the guarantee it once was. The pursuit of advanced degrees may reflect a growing reliance on credentials for entry-level jobs but does not necessarily reflect the need for greater skills in the performance of these jobs (Brooks 2005). In other words, what used to be a high school graduate's job (e.g., administrative assistant) is now a college graduate's job. As the number of college graduates has grown, more people find themselves underemployed, unable to find jobs commensurate with their education. The growing supply of college graduates has created a buyer's market in which wages are stagnant or even in decline (Perrucci and Wysong 2002). The result is that, although college graduates outearn those with less education, not all groups of college graduates experience income gains or land good jobs (Bernhardt et al. 2001). It should be pointed out that the American pattern is in some ways unusual. The pronounced emphasis on postsecondary education has not yet been replicated in Europe, although rates of university attendance are up and the demand for university graduates there is growing. However, what is common across all industrialized societies is the growth of economic inequality, indicating that the new class divides of the United States are not simply a matter of local peculiarities (F. Green 2006).

The changing relationship between the education system and work opportunity reveals the presence of serious structural lags. For example, the old economy mandated the support of public education through the high school level. High schools became the means to instill the fundamentals of reading, math, health, and social skills in future generations of workers and were primarily intended to ensure a supply of workers capable of filling the existing lower-skill jobs (Marshall and Tucker 1992). However, the fact that many of the primary labor market jobs in the new economy demand skills that far surpass basic literacy may justify public support of education well beyond high school. Ironically, as the value of education has increased, public resources directed to underprivileged students have been cut, blocking the economically disadvantaged from the opportunity to attend college (Kahlenberg 2004). But even if this support were to be provided, those resources alone would not be sufficient to eliminate the disparities that exist between those entering into the good jobs and the bad jobs. While obtaining credentials may help some individuals move into more desired employment, the secondary labor market jobs will remain. Increasing proportions of these jobs are now (and will be) filled by the college-educated.

Entry Points: Securing the Good Job in Young Adulthood

Because it is more difficult to secure good jobs with a high school education, greater proportions of young adults are delaying entry into full-time work in favor of pursuing advanced degrees. And even when they do this, they often have a difficult time obtaining a good job. For those young adults who come from families with economic resources, the difficulties are buffered, as parents can provide a variety of supports to advance the development of skills, as well as to sustain their children as they work to secure employment. For those young adults coming from disadvantaged positions, the jump from school to work is more difficult, as explained below.

Today's older workers are expected to provide more resources to their children far later in their lives than was the case for previous generations, a dynamic illustrated in Exhibit 3.5. These changes have resulted from shifts in the timing of work and family careers, a sustained redefinition of parent-child and child-parent responsibilities, and new opportunity structures that block the younger generation from independently achieving sustainable

Exhibit 3.5   A Schematic of Intergenerational Resource Transfers

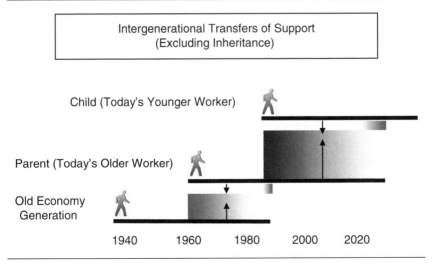

Note: In the figure, graphics indicate the following ways resources are allocated: duration (length), volume (height), intensity (shading), directionality (arrows), and timing (position).

careers as they enter adulthood. Note that the intergenerational transfer of resources can vary in the following ways: the type of resource (direct transfer of finances or provision of care work), directionality (from parent to child or child to parent), volume (how many resources are being channeled), intensity (the amount of effort involved), duration (for how long the provision of resources continues), and timing (age and life-course station). All of these means of providing support have shifted, and in the new economy parents are expected to provide more for their children, and for much longer, than was previously expected.

Historians inform us that, until the early 20th century in America (and even today in developing countries), children's roles were defined largely in terms of what they could contribute to the family economy. At very young ages, children worked side by side with their parents and commonly provided a greater share of the household income than did wives (Gratton and Moen 2004). Children also provided considerable support for parents as they reached old age. But over the course of the development of the old economy in the 20th century, children's roles were redefined as dependents. Today, this dependent status is rigorously embraced by the culture, as Americans do not expect children or young adults to channel support up to their parents (who wish not to be a "burden"); rather, parents are supposed to funnel resources down to their children (who now have come to expect this as an entitlement). As children's roles shifted to dependency, so did the cultural expectations of the types of resources parents should provide. Today childhood is viewed as a distinct life stage that should be free from the grind of life. In this context, a parent's obligation is not only to provide care, but also to give substantial supports that enable growth and exploration. While these expectations began to be established over a century ago, the quantity and quality of resources that are expected have escalated, as has the duration of provision.

Adolescence has extended itself well into ages that were previously defined as part of adulthood, and now the culture is embracing the perspective that there is an in-between stage of "emerging adulthood" that extends beyond age 18 and perhaps continues into the early 30s. These emerging adults expect freedom from parental authority, but do not assume they must be self-sufficient. They commonly rely on support from parents and take advantage of options to return home if need be. The lives of these "emergent adults" are characterized by instability in financial circumstances and interpersonal ties (Arnett 2007). With this structural change, the resources provided by parents comprise a critical safety net for younger workers. Today it is common, for example, for parents to provide support for "boomerang children" who leave home, but then come back (Goldscheider, et al. 1999).

Studies vary in their estimations of the quantity of resources that parents provide to adult children, but all conclude that substantial proportions of children in their 20s remain reliant on parents for significant housing and financial support (Arnett 2004; Gitelson and McDermott 2006; Goldscheider and Goldscheider 1999). Schoeni and Ross (2005), in an analysis of the Panel Study of Income Dynamics, found that during the transitional years (age 18–34), parents contribute (on average) about $38,000 in material assistance to each child. But this support varies considerably by socioeconomic class. Children from families in the top income quartile received $70,965 during this transition period. Aside from helping children navigate the types of unstable careers that are common in young adulthood, parental resources are also the primary source of financial aid used to fund college educations. How do parents afford this? Commonly by borrowing money, using home equity loans, or forgoing investing in their personal retirement accounts (Cha, Weagly, and Reynolds 2005). While access to parental resources helps young adults prepare for work in the new economy, deprivation of these supports significantly limits opportunities and life choices (Aquilino 2005).

In sum, substantial investments from parents are a major factor in increasing the odds that a young adult may enter into the primary labor market. For children from disadvantaged families, these resources are limited, which weighs against their career prospects (Eccles and Zarrett 2006). For parents in the middle class, the provision of these resources is commonly accompanied by increasing the pressure to be more intensive parents, to earn more, and to take on personal debt, all for the sake of helping one's children secure their careers. Not only has the structure of opportunity become more polarized, with a growing gap developing between winners and losers, but the chances of being one of those who gets a good job in the top tier of the new economy increasingly depends on parental resources and whether parents themselves have good jobs.

Is the Global Economy Becoming More Flat or Bumpy?

A major part of the reason for declining opportunities in the United States can be attributed to the loss of high-paying manufacturing jobs, the type of work that was obtainable with a limited education in the old economy. The reasons for the decline of manufacturing are complex, as discussed in Chapter 2. However, at least some of the decline can be attributed to the movement of jobs overseas. But what has happened to inequalities related to the opportunity structures in emerging economies in the global economy? Perhaps the loss of manufacturing jobs in the United States has improved the quality of life elsewhere, and perhaps the divide that separates the "developed countries" from the "developing countries" is narrowing. In this section we consider that issue, as well as how the class structure and work opportunities within developing countries have been affected by the shift to the new global economy.

There is much disagreement about the ways that the developing world is being affected by the emergence of the new global economy. At one extreme in this debate are those who suggest that globalization is creating a global race to the bottom in which employers seek out those countries in which wages are lowest and worker protections are weakest, resulting in deteriorating work conditions and economic fortunes on a global scale (e.g., Cowie 2001; Lash and Urry 1987). On the other side of the debate are those who argue that globalization is expanding fortunes and creating opportunity, both at home and abroad (e.g., Friedman 2005).

No matter how one assesses its effects, it is clear that the global economy is transforming work in developing countries, influencing workers' opportunities and behaviors in the process. One such change can be observed in accelerated urbanization in developing nations. Exhibit 3.6 reveals that in 1950, for every one person residing in a city in developing nations, there were three other people living in the countryside. By 2020, the developing world will have three times the population it did in 1950, and most of its people will be living in urban centers. One estimate is that, by 2050, more than 70% of the world's population will live in cities, and even the poorest countries of sub-Saharan Africa will be primarily urban (Saunders 2010). In some respects, the causes of urbanization are similar to those that drove this process in the United States in the latter part of the 19th century. People are establishing communities that coincide with current opportunity structures. Urban centers are expanding because of push factors (the scarcity of jobs in rural areas) as well as pull factors (the perception that jobs will be available in the cities).

As they move to urban centers, people in developing countries are transformed as workers. In a process that parallels what happened in Europe and the United States in the 18th and 19th centuries, they are proletarianized. That is, they are transformed from peasants and other nonemployees into wage workers (Deyo 1989). But in the developing world in the new economy, the problems of urbanization and proletarianization are magnified by scarcities of resources, by outmoded (and sometime nonexistent) infrastructures that operate amid contemporary technologies, and by overpopulation. Some migrants, such as Chi-Ying (the young worker discussed in Chapter 1), successfully find factory work. But many cannot, resulting in the growth of huge concentrations of marginally employed people in places such as the *favelas* of Rio and Sao Paulo or the swollen urban centers of nonindustrial Africa.

Exhibit 3.6   World Population Projections (Millions): 1950–2030

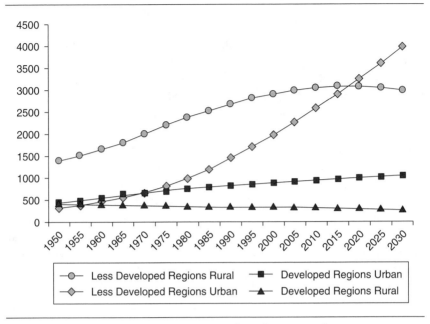

Source: United Nations, *Population Challenges and Development Goals* (2005).

Grappling with the social problems associated with urbanization—including employment, health, safety, sanitation, and housing needs—will be one of the greatest challenges for the new global economy (Cohen 2003; Economy 2005; Kim and Gottdiener 2004; Roberts 2005; York, Rosa, and Dietz 2003).

Employment prospects in the urban centers of the developing world are often grim. There simply are not enough conventional jobs available to absorb the enormous mass of city dwellers. One result has been the growth of the "informal economy" in the developing world. This refers to a variety of both legal and illegal activities, including begging, drug dealing, making and selling food, various kinds of home work, operating jitneys and pedicabs, and many other jobs that are unregulated and unprotected. Disagreement about the consequences of this type of work is intense. Some see it as positive and as opening opportunities for urban dwellers to engage in entrepreneurship that can be translated to the formal economy. These small entrepreneurial activities are seen as the basis for the gradual acquisition of wealth and, eventually, providing a springboard to middle-class status (Saunders 2010). Awarding microloans to individuals, as practiced by Nobel Peace Prize recipient Muhammad Yunus's Grameen Bank, reflects this way of thinking about cultivating work opportunity in developing societies. Others respond that the informal economy actually competes with and eliminates more desirable employment opportunities. In this manner, it is "stealth" formal economy work—as when home workers produce goods that directly or indirectly make possible the low-cost production of goods that eventually are sold in retail outlets such as Walmart and Costco (Davis 2006).

The movement of population within the developing world and the limited employment opportunities for migrants add to the pressure to move across national borders (Mattingly 1999; Sassen and Smith 1992). But even though jobs are highly mobile in the new economy, significant barriers discourage workers from moving across borders. The proportion of the world's population living outside its country of birth has only risen from 2% to 3% in the past 30 years, and, only 10% to 12% of the population of the United States and Europe is foreign-born (Glyn 2005). For some, migration is a route to economic prosperity, as the experience of many immigrants in the United States demonstrates. However, dangers exist as well. Once immigrants are removed from their local residences and social networks—and especially if they migrate illegally—they labor with fewer protections. This problem is dramatically illustrated in the lives of migrants from Indonesia who serve as domestic workers in the Middle East. These young women are socially isolated, labor for meager compensation, and receive little protection from government agents in their home or host countries. In the worst cases, they are treated more like slaves than employees (Rudnyckyj 2004).

Decentralized production and internationalization have fostered the development of what is now a hallmark of the 21st century work—**global supply chains**. In the new economy, workers seldom create commodities, start to finish, at a central location within a nation state. More often, production is network-driven, divided up between companies that are spread across the world (Schrank 2004). Some of the most recognized names in the new economy—including Liz Claiborne, the Gap, and Nike—do not directly employ any production workers. Instead, these companies develop designs and marketing strategies and then contract work with independent manufacturers operating abroad (Gereffi 1994; Tracy 1999). In this system, Nike employees can be well paid and live in good, stable communities, but the workers who produce Nike shoes often are not. The operations of these supply chains expand wealth in a progressive fashion, a dynamic illustrated in Exhibit 3.7. Relative to the profits obtained at the final stages of production and distribution, workers laboring upstream (concentrated in the developing world) produce and accumulate smaller shares of wealth (Bair and Gereffi 2003). This dynamic helps to explain the paradoxical fact that the global economy may be simultaneously expanding wealth in both the developed and developing worlds, but also expanding or extending preexisting economic disparities across the globe.

Exhibit 3.7   How Value Accumulates in a Global Supply Chain

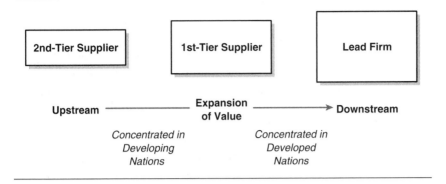

If global supply chains operate in the manner presented in Exhibit 3.7, we would expect that in the wake of the transition to the new economy, the divide that separates the rich countries from the poorer countries would remain intact, and perhaps even expand. According to a study by Milanovic (2005), a lead economist at the World Bank, this has in fact happened. Of the countries that were identified as being either in the third world (poor) or fourth world (very poor), in 1960 and 1978, nearly all remained in the same economic position in 2000, and for those that changed position, the most common direction was *downward*. And among the countries identified as being "contenders" in 1978 (who might possibly transition to the status of being a rich country—countries like Costa Rica, Mexico, and South Africa), most remained in the same position or slipped downward. From 1978 to 2000, the downward mobility rate (i.e., slipping from contender status to third-world status) of all countries was 29%, and the upward mobility rate (i.e., moving from third-world status to contender status) was only 3%. In fact, only two countries in the past 40 years moved out of the fourth world—Botswana and Egypt.

A few illustrations will help clarify how multinational organizational structures and global supply chains affect international inequalities and opportunities for individuals. We start with China, now a major force in the global economy. How powerful is its role? The answer is commonly assessed by considering its **gross domestic product (GDP)**, a figure (translatable to dollars) that indicates the total value of all goods and services produced within a country. With an annual GDP of $9.872 trillion, China now trails only the United States and the European Union in its economic weight in the world system. In 2010, approximately 1 in 7 dollars (13%) of global economic wealth was created in China (Central Intelligence Agency 2011).[2] But as we discuss next, the fact that China is catching up to the United States, Europe, and Japan as an economic power does not mean that the opportunity structures for workers are necessarily on a path to becoming like those in already affluent societies.

First, consider economic circumstances. Suppose we want to see if the typical Chinese worker is closer to the typical American worker after the transition to the new economy. How do we measure this? One method is to compare wage rates such as the **per capita gross national income (GNI)**. This is the sum of all incomes received, divided by the country's population. In 2008, China's per capita GNI was $2,940, remarkably higher than in 2000, when it was $930. In contrast, the per capita GNI in the United States in the United States was $47,930, up from $34,410 in 2000 (not adjusted for inflation) (U.S. Census Bureau 2010). China had a much higher growth *rate* than the United States, but note that the *actual* increase was substantially higher in the United States and the GNI gap between Chinese and American workers had grown. While it is impossible to predict the future, it is unlikely that the growth rate observed in China will continue to expand in a similar manner, so there is little reason to expect that the U.S.-China GNI gap will narrow any time soon. Were we to move the analysis outside of China, we would see that the per capita GNI is substantially lower in most societies outside of North America and Western Europe. For example, in Ghana, once considered to be among the most promising of the African economies, the per capita GNI was only $630 in 2008. Like Ghana, most other African countries remain poor or desperately poor, as do most countries in Asia and South America.

The fact that items and services are priced differently in different locales suggests changing the metric from per capita income (the average amount earned per person) to a measure of the purchasing power parity (PPP) exchange rates, which take into account costs of housing, food, and so on.[3] In 2008, in the United States, the per capita PPP was $46,790, in China it was $6,010, and in Ghana it was $1,320. In other words, for every one dollar a Ghanaian had to purchase needed items, the American worker had $35. For every dollar available to the average Chinese citizen, the average American had nearly $8. Will the average Chinese or Ghanaian worker "catch up" to the average U.S. worker? According to Milanovic (2005), from 1988 to 1998 the PPP increased in China, and during that time the gap that exists between the typical U.S. worker and the typical Chinese worker decreased modestly. Time will tell how far the gaps continue to narrow between China and the United States, but these types of data give a clear indication that the Chinese have a long way to go, and those in weaker economies (like Ghana) further still.

The differences in average incomes reflect the fact that workers in different parts of the world perform different types of work and are compensated differently when they perform comparable work. For example, Ohio Art's famous Etch A Sketch used to be manufactured by workers in Bryant, Ohio, who earned a livable wage before their jobs were outsourced to the Kin Ki Corporation. This toy, along with most other toys available to American consumers, is now manufactured in China by low-wage workers who are expected to work 12 hours a day, 7 days a week. These workers receive no overtime pay (even though it is required by Chinese law), have no "weekends," and are docked wages whenever they miss work (Kahn 2003). Overall standards of living in China have risen in recent years, reflecting the fact that at least some Chinese workers benefit from the new economy. However, many Chinese workers are not allowed to migrate permanently from the countryside to better-paid factory jobs. They thus become migrant workers, temporarily housed in "dormitories," and eventually are forced (often unwillingly) to return to their villages after a few years of labor (Ngai 2005). Levels of inequality within China have grown, indicating that there are both winners and losers in the process (Glyn 2005).

Industrial development in places like China has had particularly profound effects on the situation of women such as Chi-Ying. In many traditional cultures, women live in rigid patriarchal cultures where women's roles are highly circumscribed. Industrialization creates the potential for change because many of the workers who enter the new factories are women. This experience could advance gender equality, alter women's roles, and give women a degree of economic independence. The reality turns out to be more complex. On the one hand, as women become more independent, traditional forms of patriarchal domination by fathers and husbands have weakened in some ways. On the other hand, women in factories remain embedded in patriarchal kinship and community networks that restrict what they are allowed to do, and they encounter highly patriarchal work organizations in which women are the workers and men (often from their hometowns or part of their kinship network) are the supervisors (Lee 1998; Ngai 2005; Tiano 1994).

The economic gains that China experienced with its entry into the global economy also have come at a considerable environmental cost. As the land, water, and air quality in the United States became cleaner in the latter part of the 20th century (global warming effects excepted), the trend is reversed in China. Chinese rivers are now dangerously polluted, and five of China's seven major waterways are deemed unsuitable for human contact. In only six of China's 27 largest cities do residents have access to clean drinking water, and the wells of those living in rural environments have been fouled by industrial contaminants. The air is so polluted that factories had to be shut down to allow the 2008 Beijing Olympic Games to be held and airplane pilots frequently cannot see skylines when approaching Chinese cities (Economy 2005). The impact of China's industrialization goes beyond pollution. In its effort to harness waterpower for the industrial economy, the Chinese government developed numerous river dam projects that displaced millions of Chinese from their ancestral homes (Hessler 2003). Chinese economic development thus illustrates the complex effects of globalization—economic growth and rising living standards, combined with growing inequality, pollution, and community disruption. Chinese workers have reason to be both optimistic and concerned about their future.

More than low-skill manufacturing work is being directed to the developing world. India, for example, has experienced a remarkable growth in high-tech work during the past 30 years. Its successes can be attributed to a number of cultural factors, including strong work ethics and a large literate population fluent in English. The policies of the Indian government also have played an important role by directing resources to public schools, as well as to the training of engineers and medical specialists in state-supported universities. The International Institute of Information Technology (IIIT) in Hyderabad graduates 65,000 engineers and programmers a year (Bradsher 2002). As Thomas Friedman (2005) points out in *The World Is Flat*, these workers are highly skilled and are eager to perform jobs that had been concentrated in the United States.

Jobs that use information and computer technologies in the developing economies do not necessarily require workers to have technological sophistication. For example, one of the major opportunities in India is call center work. Call centers are relatively new to that society (emerging in the mid-1990s), but their entry reflects a longer-term trend to disperse service work throughout the global economy. For example, in the 1970s, companies such as IBM outsourced consumer questions to call centers in Ireland and Scotland, as well as to lower-paying rural locales in the United States (Head 2003). By employing Indians, companies shave as much as 80% off the costs of performing this same work in America and Europe (Batt et al. 2005). Call center employees tend to be young and well educated. Their work involves processing an unending stream of calls to and from Western consumers. The shifts are long (8 to 10 hours per day, 6 days a week), and the pay is low by American standards ($50 per week). And because of international time differences, most call center employees work at night. Though the jobs are clean, the work is repetitive, standardized, and intense, involving rapid transitions between conversations that last only a few minutes. The job requires the skills to bridge cultural divides and the skills to navigate the protocols of response to customer queries, tasks that can take a month or longer to master. However, these skills are not transportable to other jobs beyond the call centers and do not offer strong prospects for upward career mobility. Also note that the work hours in India are aligned to the needs of Western callers and the need to adjust to the Western languages and culture also reflect the subordinate status of these call center workers (Batt et al. 2005; Taylor and Bain 2005a, 2005b).

Other high-tech jobs in India require far greater technological sophistication. An estimated half million skilled workers labor in the Indian software industry, making it the second largest pool of software engineers in the world (Khadria 2001; Mir, Mathew, and Mir 2000). The work of computer programmers is demanding, and in India they typically work 10-hour days (and often longer). On the surface, the daily tasks of computer programmers are similar in both India and in the United States. However, the purpose of their work is often different. Indians are commonly expected to write component pieces or debug systems within larger programs designed abroad. They are seldom engaged in the creative work that results in the development of a Google or YouTube. As a result, India is unlikely to emerge as a new knowledge creation center but rather is more likely to remain an upstream link in knowledge creation, which will continue to be centered in America and other highly developed societies (Parthasarathy 2004).

The emergence of high-tech work is challenging traditional Indian customs and introducing Western cultural values. For example, traditional Indian norms largely prohibited young men and women from engaging in casual interaction, dating, and having romances in advance of arranged marriages. Call centers, with their comparatively high pay, mixed-gender workforce, and night work, are promoting sexual experimentation and increasing the number of "love marriages" among a young generation of workers. The long hours of work in the computer industries are creating the same types of stresses and overburdened schedules that American workers experience. In India, many of the strains of long schedules are buffered by the work of stay-at-home wives, as well as by grandparents who are willing and available to watch young children (McElhinney 2005; Patni 1999; Poster and Prasad 2005). But one wonders if the current generation of Indian professional women will challenge these intergenerational roles as their children have children.

Though the growth in the number of Indian programmers and engineers is remarkable, the popular image of high-tech India is somewhat misleading. In comparison with the United States, Indian high-tech workers compose an extremely small proportion of the overall workforce (there is approximately one skilled software engineer for every 2,200 Indians). Moreover, although some work in India is strongly linked to the work being performed in the Silicon Valley, the quality of life in India is still a far cry from the postindustrial experience of California (Chakravartty 2001; Patni 1999). Most Indian workers continue to engage in traditional agrarian work in rural villages or in menial jobs in urban centers; the bulk of the Indian workforce continues to subsist on extremely low incomes. Day-to-day business involves contending with power outages and political corruption. In India, like China, there exists a significant, and perhaps growing, divide between the opportunity structures and rewards characteristic of rural agrarian work and urban industrial work and those typical of the emerging high-tech sector.

So is the new economy flattening the divides that separate rich countries from poor countries or leveling the class divides that exist in developing nations? Evidence is considerable that some poorer countries are economically better off now than they were a few decades ago, but also that environmental risks have been transferred to these societies as well. While standards of life may be improving in some respects, and deteriorating in other respects, we find scant support for the proposition that poorer countries are "catching up" to the economically powerful countries. The reason for the continued "bumpiness" has much to do with the types of jobs that are distributed in the developing economies and how those differ from the types of jobs present in the advanced economies. The exception to this observation, however, can be found by considering the life chances and experiences of relatively privileged members in developing economies, whose experiences can be quite comparable to those of the more privileged members of developed economies.

Conclusion

In this chapter, we considered whether the new economy is fundamentally reshaping the distribution of economic returns received from work, and if the doors that grant access to opportunity are opening or closing. Although some have argued that the new economy is fostering a new era of opportunity at home and abroad, we caution against embracing this assumption without strong reservations. The new economy is creating affluence, but all too often in a manner that reshapes and reinforces opportunity chasms that separate those who have from those who do not. This is reflected in the changing income structures in the United States, as well as in the persistence of global inequalities. In American society, work in the new economy is shaped by two key divides. One divide separates those at the top from the rest of society. The truly affluent enjoy a radically better quality of life than does the rest of the society, and their level of economic commitment to the collectivity is weakening. The second divide separates those in the middle from those at the bottom, whose prospects for career advancement are largely blocked by the design and allocation of opportunity. In respect to the divides that exist in the global economy, the transformations that shape the opportunities at home are reshaping the opportunities abroad. Those fortunate enough to occupy good jobs in developing societies can enjoy a quality of life remarkably similar to that of the fragile middle in America. But once one considers the types of jobs being performed, the conditions under which this work is performed, the proportions of workers who are involved in "good" versus "bad" jobs, and the environmental conditions in which this work occurs, the perception of a flat world disappears. And though the successes of China and India receive considerable attention, much of the developing world has been excluded from the prospects of benefitting from the global economy. This is especially true in Africa, where incomes and life expectancies remain desperately low.

Are workers better off today than they were a few decades ago? Advocates of either an optimistic or pessimistic perspective can gather numerous examples to support their beliefs. This is possible because economies have become far more diverse than they were in the old economy, not only in terms of what is expected of workers, but also the returns that work offers. Which workers are we talking about, and in what types of jobs? Considering these dynamics reveals that multiple trajectories are at play in the new economy, trajectories influenced by social class, but also by gender, race, and nationality. In the remaining chapters of this book, we focus on the issue of work diversities and divides, both of jobs and workers, as well as the tensions that are emerging in the new economy.

Notes

1. Marginal income tax rates vary according to income levels, but the same standards are applied to all individuals irrespective of their income levels. For example, the first $100,000 that a millionaire makes would be taxed at the same rate as that of another individual who only earned $100,000. It is only the earnings obtained beyond specified thresholds that are subjected to higher tax rates. This point is commonly lost in discussions of how taxes are applied to the incomes of the affluent. Capelli (2008) has argued that declining marginal tax rates for high earners fueled the growth in CEO pay packages, as it is possible for companies to offer incentives to lure CEOs from one organization to another at costs lower than would be possible were the higher rate in effect.

2. The U.S. GDP in 2010 was $14.7 trillion; the European Union GDP was $14.9 trillion. The United States and the European Union each contributed 20% to the total world gross domestic product (Central Intelligence Agency 2011).

3. But even the PPP has problems, as the purchasing needs are not equal across countries (e.g., people in a country with a national health care entitlement do not need to purchase that service), and if an individual needs to purchase goods beyond the local level (e.g., fly from China and stay in the United States) the calculations are thrown off.

Chapter Four

Whose Jobs Are Secure?

In the latter part of 2008, the United States (and many other countries in the global economy) experienced the onset of what proved to be a prolonged recession. In the years that followed, the stock market plummeted, unemployment increased, the housing market collapsed, homes were foreclosed on, and bailouts were offered. As we write this book, Wall Street seems to be recovering. The value of stocks has increased and the kingpins of the financial sector are once again securing remarkable returns. At the same time, most of middle America remains fixed in a "jobless recovery." Approximately one in three homeowners owes more on their house than it is worth, approximately one in ten workers is unemployed, and many others have given up hope of finding jobs or are working in positions that are mismatched to their skills and needs. In this chapter, we suggest that it is a mistake to view this most recent economic downturn as an isolated event and that security will return when the "jobs recovery" hopefully happens. The evidence suggests a very different future is in store for most workers in the new economy, one in which attachment to jobs will remain more tenuous and conditional than was the case in the old economy.

Because both the nature of jobs and the needs of workers are changing in the new economy, the prospects that workers can keep jobs from week to week, month to month, or year to year are decreasing (Smith and Rubin 1997; Smith 2002; Valletta 2000). Today, insecurity has spread throughout the economy and affects a widening spectrum of workers. These include people like Jamal, the fast-food worker, who labor in traditionally unstable jobs. Perhaps Emily, the skilled contract worker, faces the greatest uncertainty that work is around the corner. Even for those with more traditional career orientations, such as Eileen, insecurity has crept into their lives as well. And Chi-Ying's opportunities (and insecurities) in a developing society are intricately connected to the declining opportunities for manufacturing workers like Dan, the former autoworker (Castells 2000; Heckscher 1996; Mattsson 2003; Milberg 2004; Neumark 2000). The technological and organizational transformations that created the new economy also have intensified various kinds of risk. As a result, working families face increasing levels of stress in keeping careers intact (Beck 2000; Hacker 2006; Kalleberg 2009; Sweet, Moen, and Meiksins 2007).

In this chapter, we focus on the issue of job insecurity and how the prospect or experience of job loss shapes workers' lives and experiences. To do this, we first consider how job and career threats were intertwined in the old economy, and how the approaches to managing these perils in the United States differed from those adopted in many European countries. We then consider workers' exposure to risk in the new economy and gauge the extent to which job insecurity has grown and why. We show that the spread of job insecurity has occurred because of the decline of older, more secure types of jobs, but also because of changing strategies for organizing work and the changing composition of the labor force. The result is that more families are grappling with economic and emotional turmoil, and they do so in a society ill-equipped to meet their needs.

Risk and Work: Historical and Comparative Views

One critical question in the sociology of work concerns who bears the burden of risk. For instance, if production orders fall, is it the employer's responsibility to continue to provide paychecks to workers? If a new technology can be developed to replace workers, is it the employer's responsibility to find obsolete workers new jobs? If an applicant is willing to work for less than an existing employee, does that employee have any right to keep that job? And in an economy built on wage labor, should citizens receive support if no jobs are available? In the wake of the Industrial Revolution in both Europe and America, the answers to all of these questions would have been no. Employers were expected to provide jobs, but at their discretion and according to terms and conditions of their own choosing, as well as with respect to what the labor market would bear. Government was not charged with the responsibility of providing assistance to those who lacked jobs or to those who could not work.

In the early phases of industrialization, workers labored without any of the protections that are present today. To secure work, they commonly had to bribe foremen for jobs, and they could expect discrimination on the basis of their ethnicity, age, and gender. If they refused to give their foremen kickbacks, or if they caused any trouble on the shop floor, they could expect to be vindictively dismissed. And if they were cast out onto the street, there were virtually no government-sponsored social supports to help them find work or survive extended joblessness (Engels [1845]1936; Jacoby 1985; Katz 1996; Thompson 1963; Trattner 1999). Even as the problems of an emerging industrial economy became increasingly apparent, U.S. politicians at the national level remained reluctant to involve themselves in workplace affairs or tackle the issue of unemployment (Lipset 1996; Skocpol 1992). In part, this inaction was the result of individualistic values deeply ingrained in American culture, beliefs that individuals are primarily responsible for their own fates and that government regulation impinges on personal liberty (Bellah et al. 1985; Galbraith 1976; Tocqueville [1836] 1969). Political inaction also reflected the control early industrialists had over political processes and their ability to discourage legislation to regulate the terms under which work was performed. In sum, in the early phases of industrialization, risk in America was primarily shouldered by individual workers.

In the United States, a dearth of government regulations and programs to protect workers created opportunities for entrepreneurs to offer risk management services in the form of **private insurance**. Though some forms of insurance had been used by industrialists (particularly those in the shipping industries), it had not been something available to workers. But by the early 20th century, workers could pay premiums to purchase health and long-term disability insurance (Jacoby 2001). During the 20th century, these insurance policies (sometimes purchased by individuals, sometimes offered as a job benefit) remained one of the central means of managing the risks associated with job loss or the prospects of disability. Today, this private insurance approach to handling risk has come to haunt the American worker because insurance companies are powerful political forces that block government-sponsored health and job-protection programs. And private insurance is becoming less accessible because rapidly rising premiums put it beyond the reach of low-wage workers and many employers.

During the 19th and 20th centuries, workers also developed collective ethnic and class-based strategies for mitigating the risks of personal hardship. Starting in the 1840s, a variety of fraternal organizations, such as the Masons and the Odd Fellows, emerged. These urban social clubs formed along ethnic lines and provided member families with small death benefits and, in some instances, health insurance. This **mutual assistance** approach to managing risk, whereby collective resources were pooled to provide support in times of hardship, was later adopted by the trade unions that formed in the second half of the 19th century. But unlike the ethnic affiliations of the early fraternities, trade unions distributed mutual assistance to members of occupational groups, thus providing skilled workers with resources that could help them survive unemployment and disability (Jacoby 2001).

As organized labor expanded its power and membership during the 20th century, collective bargaining agreements provided unionized workers with some protection from the risk of job loss. Unions fought for limits on arbitrary dismissal and victories included the implementation of seniority rights (so that veteran employees could not be laid off as easily) and grievance procedures (that discouraged vindictive attacks on individual workers) (Edwards 1979). Unions helped large portions of the workforce gain access to jobs that were well paid and stable. But union initiatives in the management of risk focused primarily on advancing their members' interests. Although some nonunion employers emulated unionized firms, many nonunionized workers (who were disproportionately women and minorities) were left lacking social supports and job protections (Jacoby 2001; Lichtenstein 2002).

In the United States, access to job security and benefits is heavily dependent on what kind of job workers have and for whom they work. Their ability to "make ends meet" requires holding a job, and, if that job is lost, finding a replacement job quickly. In contrast, the approach adopted in most European countries was to address the problem of risk by introducing entitlements—rights and resources available to all citizens independent of attachment to the labor force. These transfers from the government to individuals take a variety of forms, including assistance for unemployment, disability, sickness, old age, housing, family leave, and child care. Exhibit 4.1 reveals that only 16% of the U.S. gross domestic product is directed to public social assistance, considerably less than the percentage budgeted in most Western European countries. As a result, workers in Europe have access to a greater quantity of resources that can be used if they lose jobs, and workers in America assume far greater amounts of individual risk in comparison with workers in Europe.

The social programs common in many Western European countries are costly, and for that reason many Western European countries are struggling to generate the funds needed to support these collective resources. This problem of funding generous social programs is magnified by the aging of the workforce, which creates greater resource demands to service larger proportions of the population in retirement, and increases burdens on those who remain in the labor force. It is an open question whether these pressures will cause Western Europe to become more like the United States, or if the opposing pressures of individualized risk will compel the United States to become more like Western Europe. Likely both will happen to some degree, and the extent of change will be dependent on social pressures and the actions of interests groups.

Exhibit 4.1   Government Public Expenditure as Percentage of GDP:
International Comparisons 2005

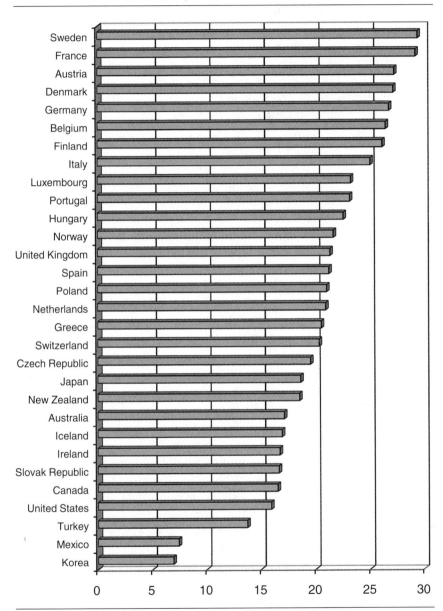

Source: Organisation for Economic Co-Operation and Development.

American workers do have some protections (see this book's appendix
for a regulatory time line). The U.S. government's first major advances
in the management of risk came as a result of the large-scale job losses

that occurred during the Great Depression. In 1935, the **Social Security Act** introduced supports for those who cannot work, including aid to the elderly, the disabled, and some children. The **Fair Labor Standards Act**, implemented in 1938, introduced many of today's most important workplace regulations, including short-term unemployment insurance and a national minimum wage. Legislative acts such as these provided vital assistance to American workers. But again, these protections are much less extensive than those enjoyed by workers in other industrialized countries.

The primary buffer against job loss for American workers is **unemployment insurance**, a program operated through federal and state government partnerships (United States Department of Labor 2007). The goal of unemployment insurance is to provide short-term assistance so that workers have the financial resources to seek suitable replacement jobs (Leana and Feldman 1995). Because the program is administered by individual states, eligibility and compensation vary within the nation. Most states provide partial wage replacement (most commonly 50% of wages—with a cap for high-income earners) for a maximum of 26 weeks. It is possible, however, for unemployed workers to receive extended benefits in times of high unemployment. But even then, this is insufficient to cover the needs of substantial portions of displaced workers. According to the Bureau of Labor Statistics, in 2011, the average length of an episode of unemployment was 21 weeks, and 44% of the unemployed were out of work for more than 27 weeks (Bureau of Labor Statistics 2011). Exhibit 4.2 shows that the support received by American workers is meager in comparison with what is available to workers in most other developed countries. As a point of contrast, Denmark provides displaced workers with 90% of their previous salary for up to 4 years.

In the United States, only one in three unemployed workers receives unemployment insurance. This is attributable, in part, to restrictions on who is eligible. Unemployment insurance generally does not cover temporary workers, the self-employed, agricultural workers, part-time workers, those who were out of the labor force for extended periods, those who voluntarily left their jobs, or those who are trying to reenter the labor force after an extended absence. It also fails to protect parents (most commonly women) who take time off to raise families. As a result, unemployment insurance covers only 40% to 60% of those who lost jobs, 10% to 15% of those who left their jobs, 25% to 35% of those trying to reenter the labor force, and only 10% of those who are attempting a new entry into the labor force. Additionally, many workers do not apply for benefits because they lack an understanding of how to apply, assume that they will find replacement work soon, or feel a sense of stigma in using social assistance programs (O'Leary and Wandner 1997; Wandner and Stettner 2000).

In sum, job-loss protections, developed during the 20th century in America, are modest, short term, and restricted, leaving workers who lose their jobs with few resources on which to fall back. This approach is

Exhibit 4.2   Unemployment Entitlements: International Comparisons

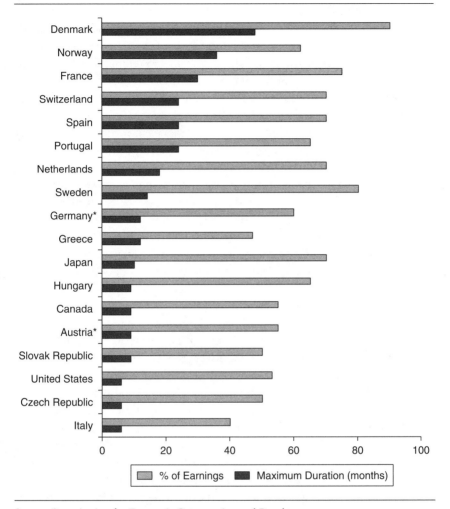

Source: Organisation for Economic Co-operation and Development.
Note: *Net earnings.

increasingly out of step with the needs of workers laboring in an economy where job loss is commonplace.

How Insecure Are Workers in the New Economy?

The advent of industrial capitalism brought with it the business cycle, with its characteristic boom and bust periods, exposing employees to the risk of job loss. However, as the old economy matured, the risk of unemployment came to be unevenly distributed. Many unionized workers successfully negotiated seniority systems that ensured that the longer one worked for an employer, the lower the likelihood of job loss. Managers and professionals were protected by reward systems designed to retain highly skilled employees and reward organizational loyalty (Cappelli 1997). Employers provided these protections not out of the goodness of their hearts, but because they were persuaded that it was in their own competitive interest to retain hard-to-replace employees and to invest in their professional development. Those at the bottom, who worked in insecure secondary labor market jobs, experienced frequent bouts of unemployment and faced high risks of destitution. As the U.S. government strove to regulate destabilizing financial speculation and to stimulate the economy when it threatened to go into recession, these countercyclical economic policies limited those risks for all workers, somewhat—especially after the Great Depression of the 1930s. Today, however, evidence is ample that something has changed, as we have seen two major bursts of very high (10% or higher) unemployment in the past three decades and it is becoming clear that even workers in "good jobs" are far less secure than in the past (Beck 2000; Kalleberg 2009; Rubin and Brody 2005; Sweet, Moen, and Meiksins 2007).

How insecure are today's workers? The answer depends on how insecurity is defined. One way of measuring insecurity is to examine the percentage of workers who are **unemployed**; that is, people who are actively looking for work but cannot find it. At any given point in 2010, nearly 20 million Americans—nearly one in 10 working-age adults—were unemployed. The rates for men were somewhat higher than for women, and African American men were unemployed at twice the rate of white men. Unemployment rates have fluctuated considerably during the past 40 years, from a low of 3.5% in 1969 to a high of 10.1% in 2010. These rates also vary monthly, within industries, and by geographic locale. For example, construction workers are commonly laid off during the winter months, but retail employment opportunities grow in advance of the holiday season.

Relying on unemployment statistics as the exclusive measure of job insecurity tells only part of the story. According to the standards used by the Bureau of Labor Statistics, to be classified as "unemployed," an individual would have to be previously employed, not currently working, *and* actively seeking a job. As a result, these statistics exclude **discouraged workers** (people who have given up trying to find work), those who have been out of the labor market and who are actively seeking to reenter it, those who are forced to accept jobs that are inappropriate to their qualifications and needs, and those laboring in jobs that offer no protection beyond short-term contracts (Cottle 2001; Kalleberg, Reskin, and Hudson 2000; Smith 2002; Thurow 1992, 1999). One illustration of the limitations of these figures can be observed by examining job creation data, also collected and reported by the Bureau of Labor Statistics. Interestingly, when new jobs are created, the numbers of unemployed do not go down proportionately (Thurow 1999). The logical conclusion is that unemployment statistics fail to recognize many potential workers, people who could be working but are not (Jahoda 1982; Kelvin and Jarrett 1985).

It should be emphasized that unemployment statistics also do a poor job of revealing the number of people who lose their jobs involuntarily. As Jacob Hacker (2006) argues, rates of involuntary job loss have grown, and the increasing rate of these types of job losses expanded well in advance of the most recent economic recession. He concluded that many people who lose jobs find replacement jobs that are inferior to their previous positions, with lower pay and benefits, and are mismatched to their education and experience. As a result, these workers remain employed, but still fear (and experience) employment insecurity (see also Kalleberg 2007). They are not, however, unemployed.

Most workers are let go one by one, in numbers too small to register as events worthy of a newspaper's lead story. But in the new economy, these trickles of job loss combine to form rivers of insecurity. Other workers are let go together with sizable numbers of other employees in plant closures. Beginning in the mid-1990s, the Bureau of Labor Statistics began tracking **mass layoffs**, events in which establishments had 50 or more employees applying for unemployment insurance within a 5-week period. Exhibit 4.3 shows that in 2010, during the "jobless recovery," at least 19,564 establishments had large-scale layoffs affecting more than 1.8 million workers over all. This was good news in comparison to the number of mass layoffs that occurred in 2009, but the number of workers affected remains high. The three most common reasons for mass layoffs were seasonal work ending, company restructuring, and permanent plant closure. Although reliable data on mass layoffs are not available for a long-term analysis, the number increased significantly in the early 1970s, particularly in the manufacturing sector, and has subsequently remained consistently high.

As the old industrial economy gave way to the new economy, decreased job security was especially noticeable among production workers in the manufacturing sector. Even unionized production workers in core industries could no longer count on the protections offered by collective bargaining agreements. Some of the most visible mass layoffs and plant closings occurred in industries (such as automobile assembly) and in states (such as Michigan and Ohio) that had traditionally been the heart of strong labor union organizations. In addition, the sharp decline in work stoppages in recent years is strong evidence that workers are less well-equipped individually or collectively to fight for improved job security or to resist employers'

Exhibit 4.3   Trends in Mass Layoffs in the United States: 1996-2010

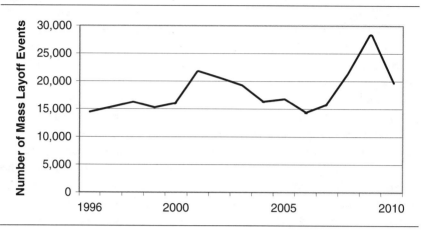

Source: Bureau of Labor Statistics.

efforts to gain "flexibility" and undercut seniority protections. Hourly workers in most industries, skilled or unskilled, unionized or not, can no longer count on the security provided by large employers and seniority rules.

The growth of job insecurity is, in part, the product of a more volatile, boom-or-bust economy, in which deregulation and efforts to shrink government's economic role have helped to fuel speculative bubbles and weaken countercyclical economic policies. However, this insecurity now also affects workers in high-skill, high-paying occupations. For example, the high-tech sector (which is often touted as one of the key elements of the emerging new economy) has been characterized by the same pattern of job loss and mass layoffs as traditional manufacturing. Even in advance of the most recent recession, Intel eliminated 10,000 jobs in 2006 and Hewlett-Packard cut 15,000 in 2005. Some estimate that between 500,000 and 800,000 high-tech jobs moved offshore in the early 21st century (Schultz 2006). Although some employers continue to place an emphasis on retaining high-skill employees (e.g., Microsoft), many employers have responded to increased economic competition and unstable market conditions by weakening their commitment to reward structures designed to encourage employee loyalty (Cappelli 1997; Moore 1996). Thus, efforts to promote "flexibility" encourage looser commitments to all employees. The flexibility push is reinforced by stockholder pressures to engage in periodic bouts of downsizing and layoffs, by new forms of work organization that assign managerial tasks to worker groups, and by new organizational structures that make it harder to advance careers by moving up within a single employer. Managerial and professional employees once could count on generous benefits packages and stable, steadily growing incomes. Today, these workers have good reason to feel less confident that these arrangements will be available and that security will expand as their careers progress. Additionally, fiscal pressure has increased insecurity among the many professionals employed in the public sector (such as teachers), and some states such as Wisconsin have attempted to limit the ability of public sector employees to collectively bargain. Though some managerial and professional workers have done well in the new economy, it is also apparent that at least some of these once-secure workers cannot be as confident that they will retain their jobs and have to think seriously about the possibility of midcareer changes (Cappelli 1999; Jacoby 1999; Moen and Roehling 2005).

Hacker (2006) notes that American society has increasingly shifted "risk" onto the shoulders of employees. Instead of defined-benefit health care plans, for example, employees are encouraged to open health care savings plans (which means they risk not saving enough). Instead of traditional pensions, employees are offered stock ownership plans, investment opportunities, and other market-based opportunities that promise greater payouts, but also involve the risk of loss. And those risks are real. This lesson was learned by Enron employees who lost nearly all their retirement savings when the company failed, because much of it was in the form of Enron stock contributed by Enron or purchased by the employees themselves. They were not alone; the 120,000 employees at United Airlines saw most of their retirement savings disappear when the company declared bankruptcy in 2004 (Marks 2004). Newspaper accounts of companies closing operations, underfunding their pensions (and petitioning the government for financial assistance), and engaging in legal maneuvering to get out of pension commitments to current and retired workers are now commonplace. By 2010, talk of modifying previously ironclad pension agreements with public sector workers was becoming widespread (Cooper and Walsh 2011).

Income stability also has weakened in recent years. There have always been people who have suffered sudden, temporary drops in income (because of illness, job loss, business failure, and the like). But the numbers of people who suffer sudden losses in income have grown significantly. Moreover, the magnitude of their income fluctuations also has grown—Americans' chance of a 50% or greater drop in income rose from just over 7% in 1970 to almost 16% in 2002.[1] And although the risks of such income fluctuations are still greater among the less educated, the risks grew more rapidly for better-educated Americans in the 1990s (Hacker 2006). With budgets stretched to the breaking point, for many dual-earner couples, an unexpected event such as a job loss or illness can plunge a family into bankruptcy (Warren and Tyagi 2003).

There is also a subjective dimension to job insecurity, the anxiety felt at the prospect that one's own job might disappear (Schmidt 2000). The General Social Survey, a widely used national opinion poll, indicates that feelings of job insecurity have been rising, especially among managers and professionals. In 1977, about one in 20 (6%) managers and professionals reported that it was either "very likely" or "fairly likely" that they would lose their jobs within the year. By 2004, the proportion had doubled, with more than one in 10 (12%) reporting that there was a serious possibility that they would lose their jobs. Another survey, the Couples and Careers Study conducted from 1998 to 2000 by the Cornell Careers Institute, surveyed dual-earner middle-class workers in upstate New York. This study revealed that most higher-level workers were only about 80% to 85% confident that their jobs would exist for the next 2 years. Only one in four (27%) men and one in three (35%) women in this study were fully certain that they could keep their jobs (Sweet, Moen, and Meiksins 2007).

Finally, job insecurity is expanding because of the changing composition of the labor force, as well as because of the changing organization of work. In an economy dominated by dual-earner families, the loss of one spouse's job creates strong prospects that the other partner—even one with a secure job—may need to reconsider career options. Those partners can be put into the position of trailing spouses and be forced to rework their careers to match the direction charted by their partner. Alternatively, displaced workers could have spouses who hold jobs that they are reluctant to relinquish. In this circumstance, displaced workers are anchored to communities, and their options can be limited. To fully understand the realities of job insecurity, therefore, requires thinking of families of jobholders as well as individual jobholders.

To highlight how this operates, let us reexamine the data from the Cornell study from the perspective of couples, and link spouses' careers together as they actually are in the lives of most Americans. Recall that the typical professional worker is about 80% confident that his or her job will remain for another couple of years. Using this as a dividing line, we can consider those at or above the 80% mark as holding comparatively "secure" jobs and those below as being insecure. Remember that these workers are married to spouses who also hold jobs that may be secure or insecure. What are the prospects of both partners in a dual-earner family holding secure positions at the same time? Exhibit 4.4 reveals a sobering state of affairs in America—*secure working conditions are the exception*, rather than the rule, for dual-earner professional couples. Only 44% of American dual-earner households have both partners feeling fairly secure in their jobs; most families have one or both partners feeling insecure. If we raise the threshold to *full* confidence that jobs are completely safe, the

Exhibit 4.4   Job Security Configurations of Dual-Earner Professional Couples (80% Confidence That Jobs Will Be There in 2 Years)

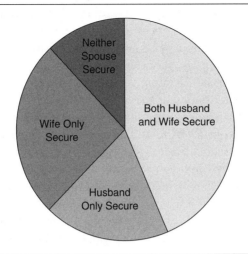

Source: Sweet, Moen, and Meiksins (2007).

numbers plummet even further, as only about one in 10 (12%) of dual-earner households has both partners feeling fully secure (Sweet, Moen, and Meiksins 2007).

Finally, we note that insecurity has been intentionally built into the design of some jobs with the purpose of maximizing the power and control of employers. This is the case in the taylorized "McJobs" in the secondary labor market, where complaints concerning work conditions can be countered with "if you don't like the job, find another one—you are replaceable." Even worse conditions are found in the employment of day laborers, who must renegotiate the terms of employment on a daily or weekly basis. Many of these laborers are undocumented workers and have no recourse other than to accept whatever terms are offered. Other day laborers use their jobs to buy drugs, and these relationships feed into a vicious cycle of substance abuse and economic marginalization, while at the same time securing the employer's ability to obtain low wage work on an "as needed" basis (Williams 2009).

The Costs of Job Loss and Insecurity

Losing a job has immediate and long-term consequences for household finances (Broman, Hamilton, and Hoffman 1990; Hoffman et al. 1991;

Leana and Feldman 1992; Sales 1995). Even among displaced professional workers, two in three receive no severance pay after being terminated. For those who do get severance packages, the compensation varies greatly. Most receive only a modest sum of money (two weeks' pay being the most common), providing little relief from the financial strains of finding new work (Sweet, Moen, and Meiksins 2007).

The economic impact of losing a job is not a new concern. What has changed in the new economy is the ability of families to weather these disruptions. In the old economy, the husband/breadwinner arrangement made it possible for wives to act as financial reserve units, as well as to hold jobs temporarily when their husbands lost work. Today, the incomes of both spouses are essential to meet family budgets. As a result, short periods of joblessness can devastate family finances in ways that were less common in the old economy (Moen and Roehling 2005; Warren and Tyagi 2003). And, this experience is becoming increasingly common. Recent estimates indicate that the percentage of dual-earner families in which one spouse lost a job doubled between 2007 and 2010, with more than a million such couples being reduced to one earner in 2010 alone (Uchitelle 2011). In our own studies, when we asked displaced workers and their spouses what they wished they had done differently, the most common responses were to have saved more and to have put more energy into finding new work before losing their old jobs. But when families are putting in long, stressful hours of work already, this is often difficult to do. As one displaced engineer, a mother of an 8-year-old boy, told us:

> When I think of myself at that time [before I lost the job] . . . I felt like I was doing just about all I could do.[2]

Another problem facing insecure workers is that even though jobs are unstable, they remain tightly tied to individuals' senses of identity and feelings of self-worth. As a result, losing a job can have a major impact on health and social-psychological well-being (Burke and Greenglass 1999; Hamilton, Broman, Hoffman, and Renner 1990; Iversen and Sabroe 1988; Jahoda 1982; Kelvin and Jarrett 1985; Leana and Feldman 1992; Perrucci, Perrucci, and Targ 1997; Shamir 1986a; Uchitelle 2006; Vosler and Page-Adams 1996). Job losses also strain marriages, sometimes to the breaking point (Perrucci 1994; Shamir 1986b; Westman, Etzion, and Danon 2001). Consider, for example, how job loss affected Lisette, a displaced 48-year-old clinical nurse, and her relationship with her husband:

> [My boss] just told me that she had some bad news. That, you know, that they had to make the budget cuts. And my position was one of the positions that was being eliminated. And, I mean, I just burst into tears. . . . And I was just so shocked. Everybody that worked with me was shocked. Just, um, it was just

really unexpected and nobody had thought that, you know, it would happen to somebody like me, you know, who had done such a good job and had worked so hard and I was really well liked by everybody there.

Feeling "lost," and perhaps in a state of denial, Lisette had difficulty understanding the permanence of her employer's decision and returned to meet with her boss in the following week to negotiate a means to get her job back. In shame, she avoided telling her husband, Damon, about the job loss and for an entire week she pretended to go to work. Damon, in turn, interpreted this as a "personal rejection" and their mutual satisfaction with their marriage dropped precipitously.

It is not only those who lose jobs and their spouses who are negatively affected. Simply witnessing coworkers being displaced increases anxieties and stress in the workplace (Armstrong-Stassen 1998; Brockner 1990; Brockner et al. 1987; Brockner, Wiesenfeld, and Martin 1995; Grunberg, Anderson-Connolly, and Greenberg 2000; Kets de Vries and Balazs 1997). However, numerous studies show that this stress can be moderated if employers make visible the rationale for organizational restructuring, provide advance notification that job loss may occur, and give evidence that procedural justice has been used to determine who will (and who will not) lose jobs (Armstrong-Stassen 1993; Latack, Kinicki, and Prussia 1995; Mishra and Spreitzer 1998; Naumann et al. 1998; Trefalt 2010; Wanberg, Bunce, and Gavin 1999; Wiesenfeld, Brockner, and Martin 1999). In other words, if workers know that jobs have to be lost, agree that job losses are necessary, and are given opportunities to plan, they are in a stronger position to adjust their lives accordingly.

Psychologists have documented how planning and a sense of control facilitate social psychological adjustment to change (Bandura 1982a, 1982b). Toward this end, advance notification can be the key to helping workers adjust *in anticipation* of job loss. If workers know ahead of time that layoffs are being planned, they are in a position to seek additional training well before they find themselves without a job. In addition, advance notice can give workers and their representatives time to negotiate with employers either to limit the numbers of people affected or to arrange for assistance of various kinds for those employees who are displaced (Moore 1996; Uchitelle 2006).

Outside of individually negotiated contracts and collective bargaining agreements secured by unions, in the United States the only "right to know" about job futures was established by the **Worker Adjustment and Retraining Notification (WARN) Act.** This law mandates that companies with more than 100 workers provide 60 days of notice if they anticipate terminating or laying off 50 or more employees. Though of significant benefit to workers in large companies, this act leaves most workers unprotected—those employed in smaller companies, those who are not part of a mass layoff, those who have worked for the company for less than 6 months, and part-time workers. Unfortunately, most employers are reluctant to extend advance notification of impending job loss. Exhibit 4.5 shows that more than half of the displaced workers we interviewed said that they received no notification, formal or informal, that their job would be eliminated. Only a small minority, one in five workers, had three or more months to prepare for the loss of their jobs. And even when employers extended severance packages, the callous methods used to inform employees of layoffs remained a bitter pill. Our in-depth interviews with managers and skilled professionals revealed sobering accounts of what it is like to lose a job in the new economy. For example, Edwin, a 40-year-old engineer and father of three young children, told us of how he lost his job:

> How did I learn my job was gone? Um, just all at once. I had no idea I was going to be let go. I was actually saying goodbye to people that were being let go. And, you know, kind of comforting them and telling them it would be okay. And I was sitting at my desk later that afternoon and I just see my boss walk up to me and ask me if I had a minute. And I kind of saw it in his face. And I say right then, "You got to be kidding." He says, "Yup, sorry, bad news." So that was it. It was that cold.

Exhibit 4.5   Percentages of Displaced Professional Workers Who Received Notification That Their Jobs Would Be Eliminated

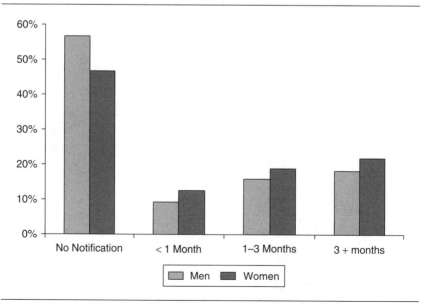

Note: N = 83.
Source: Authors' Analysis of the Couples Managing Change Study.

This case is typical. Before being let go, Edwin and his wife had it all—two good jobs, a satisfying family life, and a happy marriage. Their success came after years of working together to locate two jobs, select their community, and reconfigure their work schedules to raise their children (Moen and Sweet 2003; Pixley 2008; Sweet, Swisher, and Moen 2005). In the course of an afternoon, their lives were turned upside down, forcing them to confront difficult, fundamental decisions: how to manage dwindled resources, whose job to favor in the next stage of Edwin's career, and how to cope with the prospect of losing their home and neighborhood ties.

Even when employees have some advance knowledge of impending layoffs, the consequences can be significant. This is particularly true when layoffs affect declining regions, single-employer communities, and older workers (Sweet 2007). In these situations, workers often have great difficulty finding new jobs because little else is available locally to employ them or because of discrimination or skill mismatches. Younger workers may be willing to move away, but older workers typically are more reluctant and are more likely to experience prolonged or even permanent unemployment. The result can be both economically and psychologically devastating, undermining a worker's sense of self as a breadwinner and causing workers, both young and old, to look at work more as a way to pay the bills and less as a source of personal satisfaction and community (Koeber 2002).

Responding to Insecurity: Old and New Careers

The old economy encouraged an orientation to work and careers that is discredited in the new economy. Yet many still cling to the career mystique advanced in the old economy that loyalty would be reciprocated and that finding the good job would result in finding the good life (Moen and Roehling 2005). Today, many are starting to question this career mystique. In the old system, for those in the primary labor market, a "good employee" was someone who had found a well-paid, stable job and who displayed loyalty to the organization. Jumping from job to job tended to be seen either as opportunism or as a sign of a troubled work history. For those in the secondary labor market, unstable work histories reflected career patterns that were fostered by limited opportunity, but also were used as evidence of individuals' unsuitability for jobs that required greater responsibilities. As a result, demonstrating a lasting commitment to employers or unions became essential for upward mobility, and an employee's loyalty became synonymous with her or his "character." To prosper as an "organization man" or a "union man" required embracing an employer's or union's interests as one's own, and seeing oneself as part of an association of like-minded employees (Mills [1951] 2002; Whyte 1956). Emily, the contract worker, no longer orients her career expectations in this fashion, nor do her employers consider her character to be suspect. In fact, it is her willingness to hold short-term contracts and fulfill the work diligently that make her an attractive employee.

For those in primary labor market jobs, employee loyalties in the old economy were commonly reciprocated by employers, who adopted an ethos of **corporate paternalism**—the belief that it is the employers' responsibility to provide for their employees, much as it is parents' responsibility to provide for their children. For instance, in the 1950s, IBM CEO Thomas Watson, Jr., proudly stated, "There are many things I want this company to become, but no matter how big we become, I want this company to be known as the company that has the greatest respect for the individual." Consistent with Watson's words, IBM extended generous pay, lifetime employment, and jobs to widows (Kanter 1977). In the latter part of the 20th century, the CEO's role as the benevolent employer was replaced by a new kind of corporate leader, exemplified by Sunbeam CEO Albert "Chainsaw Al" Dunlap and the legions of corporate raiders who made quick profits by dismantling otherwise profitable enterprises (Bartlett and Steele 1992).[3] Perhaps the most vivid portrayal of the new corporate ethos is offered in the Michael Moore film *Roger and Me,* which shows General Motors CEO Roger Smith dodging the question of why American jobs were being moved offshore. In contrast to Watson, Smith defined his responsibility as being solely to the investors of General Motors.

Jobs are insecure in the new economy in part because of a growing acceptance, especially among managers and captains of industry, that jobs should be insecure (Thurow 1992). As Brian O'Reilly stated in a widely cited article in *Fortune* magazine, the "new deal" for workers is this:

> There will never be job security. You will be employed by us as long as you add value to the organization, and you are continuously responsible for finding ways to add value. In return, you have the right to demand interesting and important work, the freedom and resources to perform it well, pay that reflects your contribution, and the experience and training needed to be employable here or elsewhere. (O'Reilly 1994)

O'Reilly was writing to the select workers (and their employers) who benefited under the old system of rewarding seniority with job protections. He presented a case for the continual restructuring of work, fitting people to jobs (as opposed to the reverse), and the dismantling of jobs when work is completed. His view reflects the themes of "rightsizing" and "organizational flexibility" that were emphasized in numerous managerial publications of the past 40 years. The accepted goals of modern business are to be "lean and mean" and to restructure continually—even as a number of critics argued that this often has negative effects on corporate profitability in both the short and the long term (Grunberg, Anderson-Connolly, and Greenberg 2000; Sun and Tang 1998; Wagar 2001). Though the new social contract has been packaged as providing flexibility, from a workers' perspective it can often look much more like an effort to extend the power to displace employees at will (Pollert 1988).

The problem of job insecurity is pervasive in secondary labor markets, where workers enter jobs with virtually no protection against dismissal. For example, an application for work at a Burger King restaurant informs workers that they "can resign or be terminated . . . at any time without notice or requirement of cause." Though their job application form states that they have access to an arbitration program, there is virtually nothing to arbitrate if the signed agreement says that employers can fire workers at will (Wolkinson and Ormiston 2006).

Fast-food workers are not the only ones who experience this one-sided arrangement. Our interviews with displaced professional workers revealed many stories that show how little loyalty is extended from the top down. Even dedicated, long-service employees are displaced at will, and often in ways that show little or no appreciation for employees' past contributions. The account offered by Duke (a 36-year-old engineer at a major manufacturing company) was similar to those of many other workers who had lost their jobs:

> They had everybody in the organization go to a meeting. They handed out paper slips, some of the people went upstairs, some of the people went downstairs, and the people who went downstairs were informed that they were done as of that day, and the people who went upstairs were informed that they were going to be kept on for a couple months but their jobs would also be gone. . . . It was a real lousy way to do something like that.

As we made comparisons between workers, and looked at how employers were terminated, we found that Duke and his coworkers actually reported comparatively favorable treatment, in that they received two weeks' severance pay for each year they had been with the company, as well as temporary continuation of their health insurance and access to an outplacement company. In contrast, Tobias, a 53-year-old information specialist, recounts a worse and sadly common experience. After working a series of 12-hour days to complete an important contract, he was called into his supervisor's office, told that his job was terminated, and was escorted out of the building a few minutes later. Because of fear that he would sabotage work stored on his computer, he was not allowed to log back into the system or remove any personal communications or files on his machine. He was not alone. In the weeks that followed, 12 other employees at his company were treated in a similar manner. Tobias's story is particularly harsh, but he is among many others we interviewed who were "tapped on the shoulder" and escorted out of their buildings within a half hour of being notified that their jobs were gone.

To what extent is the new social contract reshaping workers' commitments to their jobs and employers? Some have argued that the new economy is corroding the character of American workers and undermining lasting commitments to fellow workers and employers. This portrait of the modern worker depicts her or him as pursuing immediate self-interest, but lacking a commitment to labor for a greater good. Like their employers, these workers have abandoned the values embraced in the old economy and no longer see loyalty as a virtue (Sennett 1998). Although this may be true for some employees, our interviews with displaced workers did not indicate this to be as common as is often argued. Instead of being disloyal, we found that many of the problems experienced by displaced workers resulted from their being *too loyal* to their jobs. It is possible that they did not feel that sense of connection to their employers that workers felt in the old economy, but they were intensely committed to their work and labored long hours not only because they had to, but because they saw it as their duty. For the most part, they continued to behave according to preexisting models of what a good employee should be and to give as much as they could to their employers. As a result, the most common response we heard from workers (even those who had a sense that their jobs were unstable) was that learning of their job loss was like "being a deer in the headlights," "being hit on the head," "stunned," "shocked," "distressed," "incredulous in disbelief." Had these workers constructed their value systems and personal strategies to correspond with the new economy, they would not have reported that "employers do not take care of their people," and that they felt "unappreciated," "annoyed," "miffed," "frustrated," and "extremely angry." Workers adopting a new set of work values would be much more likely to anticipate job loss and be better prepared to cope with it both practically and psychologically.

Still, workers' orientations to jobs and careers seem to be changing. The Bureau of Labor Statistics reported that in 2004 the average length of time workers had been with their current employer was just 4 years. The average person in the United States will hold nine jobs between the ages of 18 and 34, and several more during their careers. More importantly, far fewer workers expect, or plan, to stay with their current employers throughout their career, a sharp departure from the old ways of working. These data suggest that workers may be redefining career paths, as well as their strategies for navigating labor markets over the life course.

Some have argued that redesigned workplaces offer some segments of the workforce expanded possibilities to develop skills. These skills may not ensure security in internal labor markets, but they are transportable to other places of employment. This optimistic view envisions a mobile workforce charting **boundaryless careers**, engaging in transient relationships between different employers and thereby expanding a variety of portable skills (Arthur and Rousseau 1996; Raider and Burt 1996). Most analysts who adopt the phrase "boundaryless career" use it to emphasize the ways in which individual choices are expanded in the new economy. Such is the case for computer programmers in Silicon Valley. As these workers rotate through a variety of companies, they gain a variety of experiences and earn significant financial rewards, simultaneously enhancing innovation and expanding knowledge (Saxenian 1996). However, the cavalier use of the word *boundaryless* to describe careers needs to be questioned when considering the constraints on worker mobility. For some workers, especially young, educated, single workers, movement from job to job or community to community is relatively unconstrained and can result in upward mobility. But boundarylessness ignores the everyday constraints on workers that stem from their relations with others. Relocating, for most workers, entails considering how a career move may affect the career of a spouse, their children's well-being, and the ability to care for aging parents. Older or less-educated workers, in particular, find it difficult to move and are less able to acquire the new skills they might need to "job hop" successfully (Holzer et al. 2011). Rather than unconstrained careers, workers in the new economy face a variety of new constraints—in terms of social relations, economic resources, and opportunities.

The problem of career instability is also apparent in worker experiences in retraining programs. Many laid-off workers are encouraged to seek additional training, either by going back to school or by taking advantage of either publicly funded or employer-sponsored retraining programs. Workers are told that they should retool, acquire new skills, and adapt to changing patterns of opportunity. Unfortunately, worker retraining often produces cynicism and disappointment, rather than employment. It is hard to know what to retrain for, especially in an economic recession when there are few jobs of any kind to be found. But beyond this, retraining programs and supports tend to emphasize short-term skill acquisition, rather than the broad education that would make employees truly "flexible"; and retraining often winds up training people for jobs that do not exist. Displaced workers are often reluctant (quite understandably under the circumstances) to abandon their long-standing ties to occupations and industry sectors (Moore 1996; Uchitelle 2006).

Additionally, the focus on the upward mobility of unattached professionals pursuing "boundaryless careers" ignores the careers of workers who labor in less favorable arrangements. A large proportion of the workforce labors in **contingent jobs**, "as-needed" positions, filled by short-term agreements between workers and employers. These jobs include on-call work, temporary help, agency work, and contract work positions designed to be temporary and unstable (Polivka and Nardone 1989). Because of their short-term employment arrangements, contingent workers are commonly treated differently than "regular workers." For example, temporary workers are sometimes given different-colored security badges, assumed by coworkers to be less competent, not offered employer-sponsored training, and excluded from office rituals such as birthday parties. Because of their perceived marginal status, coworkers are less inclined to express interest in these workers' lives outside of the workplace and often fail even to learn their names (simply referring to them as "the temp") (Henson 1996; Smith 2002). A widely cited study found that contingent workers earn considerably less pay, have less access to health insurance, and are less likely to receive pension benefits than those who work in more conventional arrangements (Kalleberg, Reskin, and Hudson 2000). Temporary workers sometimes report liking aspects of their jobs, but as Vicki Smith (2002) forcefully demonstrates in her book *Crossing the Great Divide,* nearly all long for secure jobs, something harder to find in the new economy.

It is important to recognize, however, that not all contingent workers are alike. While many contingent workers labor in marginal temporary jobs, others operate as **contract workers**; that is, they provide services to an organization but do not work for them directly as employees. Some companies may use contract arrangements to get around labor laws that protect employees or to avoid having to pay benefits to some of their workforce. However, there also is strong evidence that many contract workers, like Emily, enjoy a sense of freedom in not attaching themselves to long-term commitments to organizations. These contract workers can sometimes command high compensation in exchange for the temporary use of their expertise, and many feel a sense of liberation by removing themselves from the confining walls of any particular organization's bureaucracy. Still, these workers also lack the conventional security offered by longer-term employment, so their security hinges on continually seeking replacement work (Bidwell and Briscoe 2009; Osnowitz 2010).

Conclusion

In this chapter, we considered risk and job insecurity for American workers in historical and international comparative contexts. We showed that the element of risk is not new and, in many respects, workers today are far better off than were workers in the early phases of industrialization. But new problems are emerging, including an expanded and broadening spectrum of jobs that offer little security. Contemporary employees face different types of risk than were experienced by employees in the old economy, not only in their employment contracts, but also in the ability of their families to weather career disruptions. And the demography of the American workforce has changed, not just jobs. Dual-earner families are in a unique position of double jeopardy because the loss of one partner's job can disrupt both partners' careers, while single-earner families (some of whom are single parent families) also face mounting risks of job loss.

Unlike their European counterparts—who enjoy some protection from risk simply by virtue of citizenship—American workers have few resources to protect them from the consequences of job insecurity and loss. American workers need to hold jobs, but also to find jobs that offer protections in the form of "benefits." When these jobs are lost, they need to find replacement work quickly. The safety nets established during the 20th century were built to shelter husband/breadwinner–wife/homemaker families through brief periods of joblessness. These risk management strategies are now woefully insufficient to protect most working families. Similarly, providing pensions and other benefits through employers has become more and more problematic, as workers are forced to move laterally from job to job and from employer to employer.

The new economy has fundamentally weakened the social contract between employers and employees. Employers once believed it was in their own interest to retain valuable employees, but economic instability, global competition, and stockholder pressure have encouraged them to move workers out of jobs at will. This new managerial ethos enables employers to serve the interests of stockholders through flexible labor management strategies, but at great cost to American workers, who often are unsure if they can depend on employment beyond the short term. And because employers seldom provide employees with solid information about planned layoffs, workers typically lack the ability to gauge their future job prospects accurately. A more equitable arrangement would necessarily involve providing workers the opportunity to plan careers proactively in an economy of expanded risk, and to make tactical career decisions based on solid information about their future employment prospects. Instead, most workers are obliged to be reactive and to redirect disrupted lives in response to decisions they had no part in making.

A return to traditional seniority-based systems of job security does not appear likely, nor would it be unambiguously a good thing. While those systems provided some workers with badly needed protection, they never covered more than a portion of the overall workforce. Moreover, questions can be also raised about the effects of these practices on labor markets and the quality of work. For example, tenure-based systems in the teaching profession have been criticized for placing younger teachers with fresh approaches (but less experience) at a decided disadvantage relative to older teachers. In an economy in which change occurs rapidly and teachers are called on to continuously retool and upgrade their knowledge and skills, employment practices that simply reward seniority are problematic. This does not mean that the elimination of employment security is a solution to motivating workers, however. What *is* needed is an employment system that rewards and encourages creative, forward-looking workers without relying simply on high levels of insecurity as a motivator.

Job insecurity will remain a contour of work in the new economy. Given this situation, we believe that discussion needs to shift from a focus on *job security* to enhancing *career security* across the life course (Moen and Sweet 2004). This involves allocating resources that help workers transition from one job to the next and eliminating the structures that force workers to be either "all in" or "all out" of the labor force. This also requires a rebalancing of the employee-employer relationship in respect to access to information, such that employees understand the prospects of their job being eliminated and discouragements introduced that curtail high labor turnover. And in an economy where roughly one in seven workers either lacks a job, or is working in a job that poorly fits their talents, attention needs to be devoted to increasing work opportunity. As we discuss in the next chapter, this may involve thinking about ways of redistributing some work from those who have jobs to those who do not.

Notes

1. These data are drawn from Hacker's (2006) analysis of data from the Panel Study of Income Dynamics, pp. 30–31.

2. Quotes in this chapter are taken from the Couples Managing Change Study (Stephen Sweet and Phyllis Moen, coprincipal investigators), which included in-depth interviews with 125 dual-earner couples who either held insecure jobs or who had recently lost jobs. For details on methods, see Sweet, Moen, and Meiksins (2007).

3. This nickname was originally coined by workers protesting his labor management practices, but was also one that he embraced.

Chapter Five

A Fair Day's Work?

The Intensity and Scheduling of Jobs in the New Economy

The workers described at the beginning of this book face a number of challenges in managing the time and intensity of their work. Eileen, the engineer, negotiated a schedule that enabled her to reduce the time she spent at work, but this contributed to her falling off the career track. To get her career back on line required her again to commit to putting in long hours of work, at considerable expense to her son's needs. Dan, the displaced autoworker, went from a job with a predictable schedule to self-employment, where he could determine his own hours. For this worker, self-employment created the freedom to fashion a schedule around his other interests, but left him feeling insecure about where the next dollar will come from. Jamal, the fast-food worker, found himself working a job with an undependable schedule that left him short on hours and income. And, as anyone who has ever worked in a fast-food restaurant knows, this kind of work provides few opportunities to rest because there is always something that needs cleaning, moving, or preparing. The job demands that Chi-Ying is expected to satisfy in her factory in China are even more intense. Like her coworkers, she commonly works 6 to 7 days per week, longer days than American workers typically work, and she is expected to do so at a pace that might best be described as frantic.

All of these workers face common problems: negotiating work arrangements that enable them to meet their family needs, keep their careers intact, and gain a fair day's wage for a fair day's labor. What is a reasonable workday for an employee in the new economy? And should employees be expected to perform, as Frederick Winslow Taylor ([1911] 1964) prescribed in his *Principles of Scientific Management,* the absolute maximum amount of labor possible? Or should their work occur at a more relaxed and (some would argue) humane pace?

This chapter focuses on the intensity and scheduling of work to understand the challenges that job expectations pose for workers and their families in the new economy. We consider the factors that cause some workers to labor longer than they want to, while (ironically) others work less than they want to, or even not at all. We consider why so many Americans feel overworked and place job demands in historical and comparative perspective to highlight the ways in which American culture and policy contribute to work-poverty and overwork. Finally, we consider the impact new schedules have on working families and their strategies for reconciling job schedules with family demands.

Time, Intensity, and Work

One obvious way to measure how much work people are performing is simply to count how many hours they spend laboring. Interestingly, this way of thinking about work commitments is relatively recent and in many ways is a very poor standard to judge the actual amount of work that is performed. Only in the past two centuries have jobs become things people "clock into." By contrast, in preindustrial times, work centered on the home and constituted a stream of activities that varied according to the rhythms of the day and the season. Workers responded to (among other things) cows returning from the field, hay in need of sowing or harvest, and children in need of nursing (Osterud 1987). The labor historian E. P. Thompson (1967) termed this *task-oriented labor,* a systemic approach that organized, measured, and compensated work on the basis of the completion of specific assignments. When work was organized according to task, schedules varied remarkably throughout the year and typically interspersed intense labor with prolonged periods of relaxation. And because work was performed in and around the home, people in preindustrial societies did not make a clear distinction between what was "work" and what was "life" because the two were inextricably intertwined.

Industrialization sparked a radical reorganization of work, shifting the criteria for compensation away from the completion of tasks toward the *length of time* engaged in tasks. This system of **time-oriented labor** led to the development of hourly wages and shift work, and it reoriented the rhythms of work and family lives around clocks and calendars (Hareven

1982; Marglin 1982; J. Thompson 1967). Although the hourly wage is now a taken-for-granted arrangement, in the early 19th century this alternate way of measuring and compensating work was not immediately embraced. Workers had to learn to arrange their schedules to match that of the factory and structure their family lives to perform shift work. As Thompson (1963) noted, the end of the Industrial Revolution can be marked by the moment workers stopped protesting over *how much* they were expected to produce and started protesting over *how long* they were expected to work—a change that transpired around 1830. The factory bells that echoed through New England in the 19th century are mostly silent, but what they were intended to accomplish continues to resonate today. Many taken-for-granted divisions of the day, week, and year—including vacations, the weekend, and the lunch hour—were all literally invented in the late 19th century and reflect the triumph of time-oriented systems of controlling work (Rybczynski 1991).

Arguably, a task orientation to work is a more natural means of organizing work than is a time orientation. Consider, for instance, that when college students are assigned to write a term paper, their most common question is "How long does it have to be?" Students never ask, "How much time will I be required to spend writing?" The first question is more natural because it concerns the product of one's efforts. The latter question is of greater concern for most workers in the labor force because their compensation is not determined directly by how productive they are, but by how long they spend engaged in the activity. Today, even some skilled knowledge work is measured by the hour. For instance, lawyers seek to maximize their "billable" hours because this constitutes the measure by which they are commonly paid (Yakura 2001). Though progress would suggest that work would have grown easier, more fulfilling, and more economically rewarding, the 60-hour workweeks of the 19th century have not faded entirely away. Many Americans find themselves working longer hours and harder than they want to, as well as more than their parents did. The hope that a new economy would enable people to work less, not more, has thus far been disappointed (Schor 1991).

Understanding time-oriented labor as a socially negotiated system of managing work presents interesting and important questions about what activities should be compensated in the new economy. Many of the problems workers confront today emerged from the ways the terms of labor came to be negotiated in the old economy. For example, one old question concerns when work begins and ends. When miners and mine owners in the late 19th century faced this question, miners ultimately were forced to agree to walk or crawl several miles to reach the coal face before they could clock in to their jobs; paid work commenced only after they had begun to dig.[1]

Exhibit 5.1   The Time Clocks That Track Worker Effort Remain a Legacy of the Old Economy

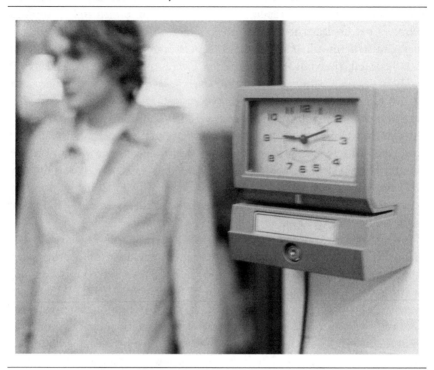

Source: Copyright © by Paul Edmondson/Corbis.

Today, this seems unreasonable, but consider this fact: The American Community Survey documented that the average worker in the United States spends 50 minutes a day (approximately 200 hours in a year) commuting to and from his or her job, but few workers are compensated for this effort. Acknowledging this, however, would require developing mechanisms for rewarding labor in all its forms, including work in and around the home, and replacing the restricted definitions of work created in response to industrialization. The fact is that many jobs today require workers to labor before they enter and after they leave the workplace. Workers in many service jobs, for instance, commonly have to perform additional work tasks, such as cleaning their uniforms, on their own time. The time and expense it takes to be educated and trained for skilled jobs in the new economy is rarely included in calculations of worker effort. As a result, many young workers enter their careers burdened with tens of thousands of dollars of debt accrued during their college educations. This is a burden that they carry, not their employers.

As we discuss later, it is not simply the number of hours worked that is the problem. Work in the new economy also is colonizing domains from which it was previously absent. Consider, for example, the extinction of the lunch hour, as well as the growing tendency to take work home on weekends, in the evenings, and along on vacations (or to forego vacations altogether). These behaviors were the exception in the 1950s; now they have become the norm. Workers find that entering careers requires earlier demonstration of significant accomplishments and larger preinvestments in training than was the case for the previous generation of workers. College faculty offer a useful case in point, as junior faculty are now expected to publish multiple articles while pursuing graduate degrees and then to jump higher bars for favorable tenure reviews than their colleagues who entered the profession in the 1970s. Nor is it only workers at the high end of the human capital spectrum who need to make preinvestments to get jobs. A high school diploma used to be the requisite for a secretarial job, but those applying for administrative assistant positions today commonly compete with others who possess associate's and bachelor's degrees. And the drive to compete is extending downward through the life course, as evidenced by the intense pressure high school students experience to "stand out" among their peers. Even then, many of these students are finding that even a near-perfect performance provides no guarantee of access to the elite college programs that would have been a lock for high achievers of the previous generation (Rimer 2007).

Measuring the extent to which work has intensified is a challenging endeavor. One approach is to consider the average hourly output of workers, which the Bureau of Labor Statistics reports have consistently advanced during the latter part of the 20th century. Interpreting these types of data is complicated, however, because some of these increases can be attributed to technological or organizational advances. But even if this is the case (as we discuss later), these innovations have not reduced the burden of work. If anything, they have increased the polarity between those who have steady work (and are intensifying their efforts) and those who do not (for whom not working is hardly "leisure"). Indeed, there is a deep irony in the fact that the technological and organizational transformations of work that increased productivity have not resulted in an expansion of leisure.

How Long Are We Working? Comparative Frameworks

The advent of a system of time-orientation at work made the issue of how *long* one worked into an important and contentious question. In the early days of industrialization, work hours were often quite long. In some cases, workers were required to put in 12-hour days on a regular basis. Accurate historical statistics are hard to come by, but economic historians estimate that work weeks of 60 hours or more were typical in U.S. manufacturing in the latter part of the 19th century (Whaples 2010). Workers and their representatives responded to long work days by pressing for reductions in work time. Many of the most famous confrontations in American labor history centered on the issue of the working day (including the famous Haymarket riot of 1886) and the effort to win a standard, 8-hour day became one of the most important rallying cries for American workers (J. Green 2006). Average work hours did decrease slowly in the early decades of the 20th century, but it was not until the passage of the Fair Labor Standards Act in 1938 that the 8-hour day became the norm in the United States. Its passage underlined what had become a central assumption shared by many Americans—that as workers became more productive, it should be possible to reduce work hours without decreasing wages.

The hope that work hours would continue to decrease as productivity grew does not appear to have been realized since the 1930s. Although there is considerable disagreement about whether average work hours in the United States are increasing or slowly decreasing, most studies show that American workers, on average, continue to work close to 40 hours per week, despite enormous productivity gains since the 1930s (Robinson and Godbey 1997; Schor 1991). Averages are informative, but equally important are statistics that document the diversity of work schedules within American society. According to the Bureau of Labor Statistics' Time Use Survey, during their workdays men spend on average 8 hours and women spend 7 hours laboring in their jobs. But as Exhibit 5.2 shows, fewer than one in two Americans worked the conventional full-time commitment of 40 hours a week. One in four Americans worked part time, and one in four worked more than full time. One startling fact is that nearly one in 10 workers commits 60 or more hours a week to their jobs. This is the group considered by pop psychologists to be workaholic, people so addicted to their work that their jobs have become their primary focus in life, which in turn causes their personal lives to suffer (Machlowitz 1980). However, ethnographic accounts of lower working-class lives reveal that long hours often are not a choice, or the result of an "addiction" to the joys of cleaning floors and waiting tables. People in jobs like these work long hours because it is the only way to make ends meet (Ehrenreich 2001). Those in "good jobs," those working as managers and professionals, report that they would prefer to work fewer hours. Surveys show that a sizable proportion of the workforce (as many as one in five) would be willing to take a reduction in pay if it would enable them to work shorter hours (Galinsky and Bond 1998; Golden 2005). When highly committed workers are asked why they work longer than their ideal, the most common reason they offer is that "my job requires it" (Moen and Sweet 2003).

Exhibit 5.2   Average Hours Worked: United States 2009

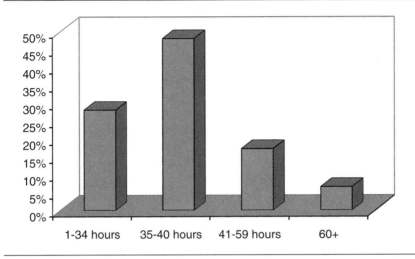

Source: Statistical Abstracts of the United States.

The problem of overwork takes on a different and more significant meaning if we look beyond statistics about individuals. Analysis of data from the Current Population Survey (CPS) reveals that men's average work hours have not changed much during the past 30 years. However, what has changed is the likelihood that there will be *two* paid workers in the family and that women will be holding full-time jobs. These facts call for shifting the analysis of work hours to focus on households. Jacobs and Gerson (2004) reveal in *The Time Divide* that the single biggest explanation for the expansion of work hours is the increasing prevalence of dual-earner couples.

Exhibit 5.3 illustrates the impact this has had on working families, showing the changes that occurred from 1970 to 2000 in the hours worked by husbands, wives, and couples. In 1975, the average wife worked only 704 hours (13 hours per week), but by 2009, these hours had nearly doubled.[2] The increase in women's paid employment was not associated with a decline in men's work hours, which also increased slightly during this period. As a result, in comparison with 1970, the "average" couple now puts in nearly 10 more hours into paid work each week. During this same period, there was little change in the total time spent on "home work," which in the 1980s (27 hours) remained approximately the same as in 1965 (24 hours) (Robinson and Godbey 1997; Schor 1991). As a result, considerable strains are created by dual-earner work arrangements, which often pit work commitments against family obligations, leaving workers—especially working mothers—stressed and exhausted (Hochschild and Machung 1989; Moen and Roehling 2005; Pitt-Catsouphes, Kossek, and Sweet 2006).

Exhibit 5.3   Trends in Husbands', Wives', and Couples' Combined Weekly
Hours of Paid Work: United States 1975–2009

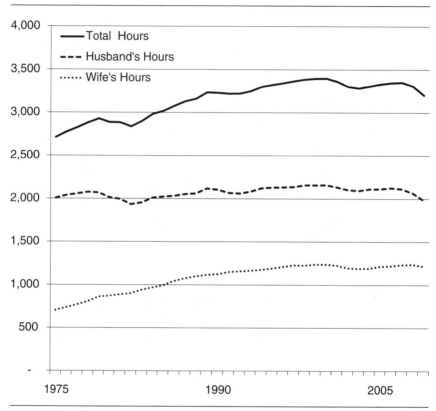

Source: Economic Policy Institute Analysis of Current Population Survey, Annual Social and Economic Supplement Data.

Another means of gauging American overwork is to examine how much time is spent working in other countries. Comparisons reveal that people in only a few countries labor as many hours as workers do in the United States. Exhibit 5.4 shows that Americans work, on average, 54 hours more per year than Japanese workers (roughly 1.5 extra weeks), 122 hours more than British workers (roughly 3 extra weeks), 214 hours more than French workers (roughly 4.5 extra weeks), and 390 hours more than workers in the Netherlands (roughly 10 extra weeks). Although Mexico, Greece, and Poland all report higher average work hours than the United States, women in these countries are less likely to be in the labor force. Thus, the United States is exceptional in the amount of time its population devotes to work.

In comparison with most other countries, the United States is unusual in its lack of policies to limit the extent to which employers can overwork

Exhibit 5.4   Average Annual Work Hours: International Comparisons 2009

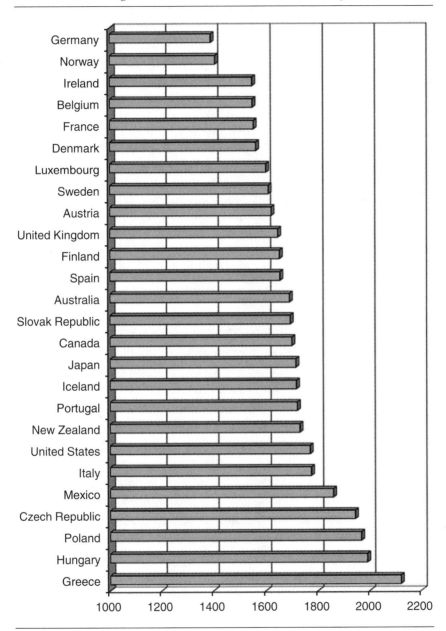

Source: OECD 2011 Online Employment Database.

their employees. Two in three countries in the world have established laws to regulate the maximum number of hours employees can be expected to work (most commonly between 48 and 60 hours per week). Employees in most countries have a right to rest breaks, which in some countries are taken collectively, thus enabling nearly the entire society to pause from work. In the United States, no such laws exist at the federal level. In fact, laws and corporate cultures in the United States reinforce the *employer's* right to demand that workers engage in overtime work, even when that work directly conflicts with family needs (Jacobs and Gerson 2004; Wolkinson and Ormiston 2006).

The United States has fewer public holidays than most other developed countries and has no law guaranteeing its workers a right to vacations (McCann 2005). Exhibit 5.5 reveals that most European nations mandate that employers give workers 3 to 6 weeks of paid vacation a year. In contrast, the typical full-time worker in the United States, though not entitled by law, gets about 2 weeks of vacation per year. Unionized employees receive more vacation, but with declining numbers of employees covered by collective bargaining agreements, the ability to secure vacation time has

Exhibit 5.5   Statutory Annual Paid Vacation Days: International Comparisons

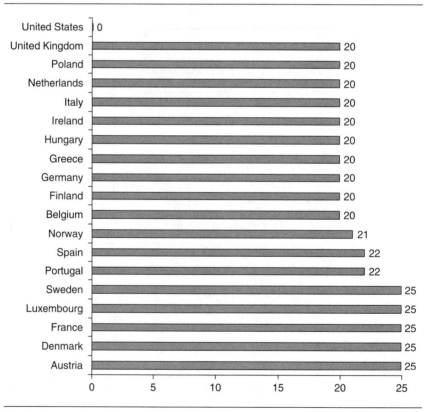

Source: Mishel, Bernstein, and Allegretto (2005).

eroded (Mischel, Bernstein, and Allegretto 2005). Additionally, American workers fail to use all of their vacation time, or they take work with them on vacation, for fear that they will fall behind in their jobs or careers.

Interestingly, Americans and Europeans took similar amounts of vacation in the 1930s, 1940s, and early 1950s. The subsequent divergence occurred because of different approaches to implementing vacation policy. In the United States, vacation leave was introduced in the 1930s and 1940s by employers, who regarded it as a means of increasing productivity by increasing employee satisfaction and health. It also was included in collective bargaining agreements, which expanded the amount of vacation time available to union workers. By contrast, most European countries introduced vacation time as part of government policy, along with a host of other entitlements such as universal health care and job protections. Because of this, vacations are available to all workers in these countries, not just workers who have jobs that offer benefits or those protected by union contracts. If similar mandatory vacation laws were enacted in the United States, American workers would be laboring closer to the number of hours worked by Europeans (Altonji and Oldham 2003).[3]

In sum, many workers in the United States are laboring long hours and families are working more than they did a few decades ago. They work more than workers in other developed nations, and often more than they want to. It is also important to recognize, however, that many American workers lack jobs or would like to work more (or more steadily) than they currently do but have trouble finding enough work. In contrast to workers who work long hours, those who work short hours are likely to say they would like to work more and wish for more predictability in their schedules (Lambert 2010). Some part-time workers voluntarily are working reduced schedules, but the numbers of involuntary part-time workers have been growing. A recent study found that the percentage of total employment made up of involuntary part-time workers had almost doubled between 2006 and 2008, accounting for 5.1% of all workers and a total of more than 7 million people (U.S. Bureau of Labor Statistics 2008). If we add that there are more than a million temporary workers in the United States (at least some of whom would prefer steady employment), over 7 million people working multiple jobs, 25% or more of whom are combining multiple part-time jobs, and unemployment rates that have reached 10% or more in the recent economic downturn, the reality of a lack sufficient work becomes even more clear (U.S. Bureau of Labor Statistics 2008). During the 20th century, a divided economy emerged, one in which there is a work-rich labor force that labors long hours and a work-poor labor force that lacks sufficient work opportunities and compensation. Arrangements that compel some workers to perform the jobs of two workers (a product of both the number of hours and the intensity of work) block the creation of opportunities that could lead to employment for all. The problems of overwork and underemployment, therefore, need to be viewed as two sides of the same coin (Gorz 1982; Jacobs and Gerson 2004; Rifkin 2004).

Finally, it is also important to remember that the reconfiguration of work demands is affecting developing societies as well. In factories in China, and in the maquiladoras that line the Mexican border with the United States, workers commonly labor 12-hour days throughout the entire week. As factory work has expanded in these industrializing regions, rural migrants find themselves subject to the rigid, clock-driven rhythms of time-oriented labor. Although these labor practices make it possible for Americans to purchase inexpensive commodities for use in their scarce leisure time, they are a burden placed on other workers such as Chi-Ying. It is hard not to conclude that the new economy, in both its domestic operations and global operations, is compelling workers to labor more, not less, and expanding the chasms that separate workers from the opportunity to have a life off the job.

Working Long, Working Hard

Americans work long hours and work very hard when they are at work. Even some workers whose hours are not unusually long, and whose hours of work have not increased over time, complain of feeling overworked (Robinson and Godbey 1997). Research conducted in 2011 by the American Psychological Association found that more than a third of Americans reported that they were "stressed out" at work on a typical work day and that "too heavy a work load" and "long hours" were two of the top five sources of stress for American workers (American Psychological Association 2011). Findings such as these indicate a collective belief that the pace and intensity of labor and the duration of work are increasing, a view that most scholarly research supports.

Jamal's low-skill restaurant job, like other forms of work designed in the old economy, were structured to pressure workers to "hustle" and "bull and jam." As many of these jobs remain in the new economy, new pressures have been introduced by new workplace practices and opportunity structures. For example, although smart technologies have eliminated some of the unpleasant, physically demanding tasks of the past, they also pressure workers to keep up with machines that work steadily and fast. Even workers whose jobs involve monitoring technology can experience high levels of stress because this type of work often involves a combination of repetitive tasks with the need to stay alert. To get a sense of these pressures, consider what it is like to be a baggage screener or ticket agent at an airport.

In contrast to the collective pressure exerted in the old economy to restrict production and adhere to the "stint" that limits the amount of work everyone will perform, the new economy appears to operate with the opposite pressure to "keep up" with everyone else (Burawoy 1979; Hochschild 1997; Shih 2004). One reason for this shift has been the introduction of managerial approaches modeled after those adopted in Japan. These systems create work teams that are structured so that each member of the group becomes a supervisor for everyone else, creating a collective peer pressure to work harder (Parker and Slaughter 1988). This type of informal social control may be proving itself to be more powerful in extracting labor than the older managerial systems that relied on supervisors "driving" workers within a top-down system of control.

Another aspect of contemporary work that leads to intensified effort is the need to do many things at once. Jobs in the new economy require workers to "wear many hats" and to have the capacity to "multitask." Time management and prioritization have become mantras of contemporary human resources departments, not just because workers have too little time, but also because they have so many (different) things to do. In addition, workers who are expected to perform many tasks are less likely to find themselves with "down time" on their hands—an employee with nothing to do is likely to be reassigned to one of the several other tasks he or she is capable of performing. Compounding the problem further is how new ways of working blur work-family boundaries. For example, computers can open avenues for workers to labor while at home (e.g., respond to e-mails) and to connect with family while at work. At the same time, computers may make it more difficult to disengage from work roles. As the anthropologist Charles Darrah (2006) observed, these time-strapped families do not "balance" work and family; their lives are endlessly busy.

Finally, contemporary workers are often pressured to labor with intensity because employers deliberately choose to limit staffing levels to keep their companies "lean and mean." The result of these efforts to reengineer and downsize corporate workforces is a smaller number of workers carrying out more work. And in the context of pervasive job insecurity, economic recession, and weakened union power, individual workers feel intense pressure to work hard, lest the axe fall on their necks next. A poignant example is offered by American air traffic controllers, whose union was shattered during Ronald Reagan's presidency, and whose growing workload has been a subject of controversy ever since (Vaughan 2006). Whether these changes contributed to recent mishaps (such as controllers falling asleep on duty) is open to speculation.

Why Are Americans Working So Much?

A variety of factors contribute to Americans working longer hours than workers in most other countries, but chief among these is economic need. For those workers at the bottom of the opportunity divide, working long hours is a means of making ends meet. Consider the fact that the 2011 federal minimum wage rate is $7.25 per hour and that its purchasing power remains well below what it was at its peak in the late 1960s, despite recent increases. A full-time minimum wage weekly paycheck of $290 is barely sufficient to provide for individual needs, let alone enable a worker to raise a family. So some workers work multiple jobs or willingly work beyond the 40-hour norm. It is also important to recognize that many minimum wage workers are employed as part-time workers but work for multiple employers, thus putting in full-time (or often greater than full-time) commitments with no prospects for overtime pay. Economic considerations also contribute to the overworking of individuals who hold managerial and professional jobs, but for these workers, career considerations or the threat of job loss is the motivating factor. Workers in such jobs fear that resisting pressures to work long hours will be interpreted as a lack of commitment, hurt their chances for promotion, or mark them as expendable when cuts are to be made (Fried 1998).

Beyond economic factors, American workplace cultures play an important role in pressuring workers to labor long hours. One of the most important studies to reveal this dynamic is Arlie Hochschild's (1997) *The Time Bind*. This book examined the work and family lives of professionals employed at Amerco, a company that offered numerous family-friendly programs, including options for reduced hours and family leave. To her surprise, Hochschild observed that even when Amerco introduced family-responsive policies, few workers used them. This raises an interesting question: if some American workers have access to leaves of absence or reduced hours, why do they not take greater advantage of existing opportunities to work less? The answer can be found in organizational cultures that encourage Americans to see long hours of work as evidence of personal competency (Blair-Loy 2003; Fried 1998; Shih 2004). Workers who occupy "good jobs" in the new economy labor with peers and supervisors who distribute badges of honor to those who can put in long hours and stigmatize those who do not. In companies like Amerco, workers who failed to put in long hours were viewed by their coworkers and supervisors as "nonplayers" who lack the qualities needed to move up the corporate ladder. The prospect of falling off the career track kept many working longer than they wanted, but others recognized that a choice to scale back could reduce their jobs to the least interesting and rewarding tasks (see Barnett and Gareis 2000).

The Time Bind also offered a radical new interpretation that inverted the "home is a haven in a heartless world" thesis, concluding that many workers in the new economy find their lives enhanced in the workplace, rather than in the family. One of the subjects in the study explained why she gladly puts in overtime:

> I walk in the door and the minute I turn the key in the lock my older daughter is there. Granted, she needs somebody to talk to about her day. . . . The baby is still up. She should have been in bed two hours ago and that upsets me. The dishes are piled high in the sink. My daughter comes right up to the door and complains about anything her stepfather said or did, and she wants to talk about her job. My husband is in the other room hollering to my daughter "Tracy, I don't ever get any time to talk to your mother, because you're always monopolizing her time before I even get a chance!" They all come at me at once. *(p. 37)*

In contrast to the romantic ideals of family life, *The Time Bind* argued that time-strapped Amerco workers experienced home lives of whining children, nagging spouses, and an unending backlog of onerous household chores. By contrast, their jobs offered friendly coworkers, concrete rewards, and interesting tasks, leading workers to *prefer* to be in the office rather than in the home.

Hochschild's study drew attention to two crucial aspects of contemporary American life: the difficulties of home life in families where both parents work and the fact that many workplaces offer real rewards to workers willing to put in long hours. Nevertheless, Hochschild can be criticized for overstating the degree to which American workers are *voluntarily* choosing work over home. Even some of her own interview subjects indicated that they worked long hours because they *had* to (Meiksins 1998). It is also important to remember that the type of worker studied at Amerco tended to be from the creative class, representing the values and perspectives of workers whose jobs offered numerous intrinsic and financial rewards (Florida 2002). The lives of these employees stand in stark contrast to the legions of workers, such as maids, fast-food workers, or retail clerks, who labor in backbreaking jobs or in hostile work environments. Even within its analysis of the creative class, the Amerco study failed to consider other professional workers who engage in creative work but who are exploited by new organizational structures. For example, colleges and universities employ increasing numbers of part-time adjunct faculty, who commonly work long hours, often at multiple jobs, for a fraction of the pay commanded by their full-time counterparts.

Long hours in the workplace offer opportunities to earn badges of honor for American workers, but this cultural orientation to work is less common in Europe. For example, Americans take pride in broadcasting their heavy schedules to coworkers to increase their status. In contrast, when German workers put in long hours, their colleagues view them as inefficient and wonder what is slowing them down. Thus, cultural differences may account for some of the disparities between American workers and European workers. But there are structural reasons as well. Interestingly, while Americans have been increasing their work hours, European nations have experimented with shorter workweeks. Most notably, during the 1990s, the French government passed legislation discouraging French companies from employing workers beyond 35 hours a week (Fagnani and Letablier 2004). This legislation reflected a desire to reduce the problem of overwork, to translate productivity gains into increased leisure, and to spread the available work around to all members of the society.

What are the consequences of legislating shorter work hours? Many European employers (and some European governments) fear that overall economic productivity and competitiveness is being harmed. Voices have been heard in many European countries (notably in Germany and France) calling for a relaxation of limits on the workweek and a move to a more American-style approach to employment, but this is not favored by organized labor or most workers. In 2008, arguing that French employment law was hobbling French business, the more conservative government headed by Nicholas Sarkozy passed changes significantly weakening the 35-hour rule and allowing longer work weeks for many workers. Nevertheless, the 35-hour week remains the reference point for French workers, and remains very popular; French workers who labor beyond 35 hours are entitled to extra pay beyond their normal hourly wage (Bennhold 2008). Elsewhere, counterarguments and proposals have been made, and there has been an effort to "standardize" work hours across Europe by establishing a 48-hour ceiling for how many hours a worker can be required to work per week in European Union countries. This would force some countries, such as Great Britain (which thus far has opted out), to alter their current practices. The controversy continues and, so far, European work hours appear to be continuing their gradual decline, indicating that the U.S. Europe gap in work hours is unlikely to disappear (European Foundation for the Improvement of Living and Working Conditions 2011; Meiksins and Whalley 2004).

Another reason for overwork in the United States is that Americans face greater penalties than Europeans if they choose to reduce their work hours. For example, most working families in the United States secure medical care as a benefit from working a full-time job. In contrast, citizens of many European countries can cut back on work, or exit the labor force entirely, and still retain access to medical care (Meiksins and Whalley

2004). Similarly, most other developed nations have invested in extensive public transportation systems, infrastructures that afford residents the option not to purchase automobiles. In contrast, the minimal public transportation infrastructure in the United States builds in an expectation that Americans purchase cars and pay for their upkeep (Feagin and Parker 1990).[4] Consider, too, the impact of educational funding in the United States, where the system ties school quality to local property values. This system entrenches social inequalities and propels economic competition to purchase property in desired school districts (Kozol 2006; Marshall and Tucker 1992).

Finally, overwork is structured into the American system of allocating jobs. To understand how this happened, we need to go back to the Great Depression, an era in American history remembered as the time when many citizens lacked jobs. Less commonly recognized is that during the Great Depression, those who did have jobs commonly worked very long hours. To resolve this dilemma, the Roosevelt administration introduced overtime provisions as part of the Fair Labor Standards Act (FLSA). This law introduced a financial penalty for overworking employees, requiring employers to pay time-and-a-half for each additional hour worked beyond 40 hours per week. This legislation helped redistribute work opportunities from those laboring too many hours to those not working at all (Costa 2000). But because of the high costs of health insurance and other employer-sponsored benefits, it is now far less expensive to expect a smaller number of full-time employees to work overtime than to hire a larger number of full-time workers to work a 40-hour week (Tilly 1995). These expenses also increase incentives to create part-time or temporary jobs that are not eligible for benefits. When employers face pressures to increase production, they tend to do so by asking their current workforce to labor more or by hiring workers to whom they do not have to pay benefits.

The FLSA defined two classes of workers: hourly workers who are entitled to overtime pay, and "exempt" salaried workers (most commonly employed in managerial or professional jobs) whose work hours are not regulated. These exempt employees are among the most time-strapped in the new economy, having experienced an incremental expansion in job expectations during the past century. As Jacobs and Gerson (2004) note in *The Time Divide*, the exemption of some workers from overtime rights has increased the incentives to overwork salaried employees and to push more workers into jobs designed to have fewer protections. One illustration of this process can be found in the ways job boundaries can be blurred in the new economy. Such is the case for "team leaders" on production crews—workers who are primarily engaged in production work, but who also have some supervisory responsibilities. How should these employees be classified—as hourly workers with rights to overtime pay, or as salaried workers who have no such rights? In 2004, the Bush administration passed legislation that excluded some hourly workers who performed supervisory work, as well as those achieving higher incomes, from overtime eligibility. The Economic Policy Institute estimated that this simple act of redefinition, which redrew the boundaries of exemption from the FLSA, resulted in 8 million fewer workers being entitled to overtime pay. It also created incentives for employers to shift hourly workers into salaried "management" positions that, in reality, have few managerial responsibilities. What changed were not the day-to-day tasks, but rather, the right to extra compensation when laboring beyond the 40-hour threshold (Eisenbrey and Bernstein 2003).

In sum, why are Americans working so long and hard? Part of the answer rests in economic need—reasonable hours for those at the bottom do not provide a livable income, and reasonable hours for those in the fragile middle do not secure a stable career. Other answers can be found in an American culture that praises those who overwork, as well as policies that fail to discourage overworking employees.

Nonstandard Schedules: Jobs in a 24/7 Economy

The old Dolly Parton song "9 to 5"—the theme for the film of the same name—included the exasperated statement, "What a way to make a living." Though many working women (and men) still labor in the traditional 40-hour-a-week grind, they compose a far smaller proportion of the workforce. Workers in the new economy adhere to a diversity of schedules, including part time, split shifts, full time, and overtime. Fewer work by the hour and more are employed by the contract or in salaried positions that demand 50 to 70 hours of labor per week (or more). Workers labor during the day, but also in the evening, at night, and on the weekend. Some clock in and out of their jobs, but other employees are placed "on call" and can be summoned to work at a moment's notice. While some workers still labor according to traditional schedules, more are demanding flexible arrangements. For still others, flexible schedules are designed by employers to meet their own needs.

The reorganization of work, allowing it to be performed around the clock, and just in time, is one of the contours of work in the new economy. Of course, working outside of "standard" hours is not an entirely new phenomenon. Even in the old economy, certain kinds of workers could expect to work nonstandard schedules. Industrial workers, in particular, typically worked "shifts," and those shifts might involve early morning starts or late ending times, or might even involve night shifts, especially when production was booming and a third shift was laid on. Similarly, safety workers such as police officers or fire fighters often worked nights, and employees in the limited numbers of businesses that were open at night (e.g., restaurants and entertainment venues) worked unusual hours. Still, until recent decades most stores and businesses were closed in the evenings and on Sundays and the 9-to-5 schedule was perceived as the norm for workers outside manufacturing (and day shifts were the norm even for industrial workers).

The new economy has made the idea of a "standard" work day or week problematic, however. It offers no week "ends." It continues 24 hours a day, 7 days a week. In the United States, no Sabbath, blue Sunday, or holiday compels workers to take a break. Notice, for example, how some stores proudly announce that they are open "24 hours a day, 365 days a year." Increasingly, some workers find that they are being "asked" to work on national holidays, such as New Year's Day or Labor Day. As the demand for workers to fill jobs around the clock increases, so too do the challenges faced by working families. The elaborate family schedules (typically posted in kitchens) that reconcile conflicting obligations of working parents and children attest to the complexities of melding work and family schedules. Just as the Industrial Revolution forced the culture to adjust to new rhythms, the emergence of a 24/7 economy is creating profound changes in work and family lives, as well as in other institutional arrangements. This presents new opportunities for working in new ways, but also new challenges in adjusting lives to nonstandard work schedules.

Exhibit 5.6 illustrates one means of capturing the extent to which nonstandard work schedules are common in the new economy. While it is true that the daytime schedule is the time when most workers are laboring, note that even at 3 a.m. approximately 5% of workers are at their jobs, and at 6 a.m. or 9 p.m., 15% to 16% are working. According to research by Harriet Presser, the leading expert on nonstandard work hours, when the number of work hours and the timing at which work occurs are taken into account, fewer than one in two employees (40%) works a standard day shift approximating the notion of a 9-to-5 full-time job (Presser 2003). One recent study of younger workers in the early 21st century (between the ages of 21 and 39) found that 80% of such workers had had at least some experience of nonstandard work schedules, and that this was true for both men and women, and for members of all racial groups (Presser 2008).

Exhibit 5.6   Percentages of Employed Persons Working on Their Main Job at Different Hours of the Day and Night: 2005–2009

Source: Bureau of Labor Statistics American Time Use Survey 2005–2009.

Part of the reason for the expansion of nonstandard scheduling is the emergence of new markets. For instance, evenings and weekends present lucrative opportunities for retailers to increase sales. The same is true for employers in the fast-food industry, which in the 1980s expanded business hours to enable the introduction of breakfast menus. The importance of offering shopping hours at nonstandard times has grown with the increasing prevalence of dual-earner couples. Few families can function if stores and other businesses are open only between 9 and 5, and many time-strapped families rely on restaurants and other commercial outlets for services they once performed themselves. Many of the jobs created by the expansion of work time beyond 9 to 5 are part-time jobs, as evidenced by the fact that part-time workers have been found to be most likely to work nonstandard shifts outside core daytime hours (Beers 2000).

New managerial practices and philosophies also increase pressure on manufacturers to extend work into the nights and weekends, thus contributing to the creation of nonstandard schedules. Flexible production methods discourage employers from keeping a large stock of component parts ready for assembly, instead favoring the just-in-time approach, in which parts are manufactured (and often received) close to the time at which they will be needed. This new approach helped companies such as Harley-Davidson modify product lines with greater ease, minimize persistent quality problems, and create products tailored to individual customer desires

(Reid 1989). But producing goods "just in time" requires the creation of demanding production schedules that may operate into the evenings and weekends. In the wake of expanded markets and reconfigured production methods, distribution companies such as UPS have expanded their operations, employing legions of workers to labor throughout the night processing packages, enabling a lobster caught in Nova Scotia to be transported live to a restaurant in Kansas (McPhee 2005).

Why do people work nonstandard shifts? One reason is linked to the desire for part-time work. For example students who need to work before or after school or workers who are trying to work more than one job may find nonstandard shifts a useful expedient. But, part of the answer also can be found in job markets and the types of opportunities available to workers. For those in the lower classes, for whom well-paid work is more difficult to find, the need to find any work can require bending one's schedule to fit the needs of employers. Survey data indicate that the most common reasons for working odd hours are that they are built into job definitions and that opportunities for working a regular shift are not available. The next most common reason offered—and this is more often the case for women—is that nonstandard job schedules offer greater opportunity to manage personal or family needs, such as helping with child care or enabling workers to continue their education (Presser 2003; Wharton 1994). It is not unusual for spouses to engage in a tag-team strategy of managing family—as one spouse returns from work to home, the other spouse leaves the home to go to work. But this adaptive strategy comes at a cost, as spouses' lives are adversely affected by their short contact with each other, and as the bonds that unite couples weaken in the absence of rituals such as the evening dinner (Ochs et al. 2006). An award-winning study revealed that recently married couples with young children who adopted alternating schedules were *six times more likely to get divorced* than were similar couples who worked regular schedules (Presser 2000).

How Americans Deal With Overwork

Having a good job is often accompanied by the expectation that one will labor long hours, uninterrupted, year in and year out. Those who hold bad jobs, many of which are located in the expanding service sector, need to labor extremely long hours to afford even a scaled-down version of the American dream. Foregoing work comes at great costs, and many American workers face a stark choice: accept long hours as a way of life or choose to work shorter hours for considerably less compensation or fall off the career track. As a result, the most common response of workers is not to challenge employers or government leaders to change workplace policy, and instead bend their lives to fit within existing structures (Moen and Roehling 2005).

Younger, childless workers are more apt to be willing to put in long hours, whereas those later in the life course, who have caretaking responsibilities, or whose health may be failing, experience greater difficulty adjusting their lives to fit the expectations of employers. Interestingly, most young men and women do not plan to make radical changes in their work hours following the birth of a child. But after trying to "have it all," many find the demands of work and family too overwhelming to manage. Faced with mounting pressure, the most common response is to fall back on gender roles developed in the old economy (Becker and Moen 1999; Stone 1997).

One way of capturing this dynamic is by studying the different schedules dual-earner couples adopt before, and after, they have children. Exhibit 5.7 shows the work-hour arrangements adopted by two groups of dual-earner couples, those who are younger nonparents (dual incomes, no children), and those who have a young child. Five work-hour arrangements are identified:

- High-Commitment Couples (both spouses working long hours—more than 45 hours a week)
- Dual Moderate Couples (both working full time—35 to 45 hours a week)
- Neotraditionalists (husband working long hours but the wife not)
- Crossover-Commitment Couples (wife working long hours but the husband not)
- Alternate-Commitment Couples (neither working long hours and one working reduced hours)

Exhibit 5.7   Work-Hour Arrangements of American Middle-Class Dual-Earner Couples

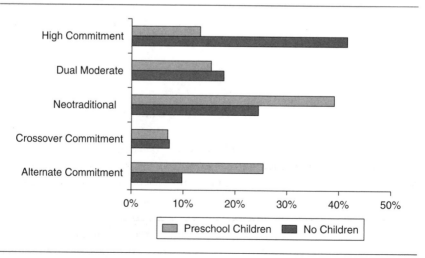

Source: Adapted from Moen and Sweet (2003).

Notice that in the new economy, the high-commitment arrangement (with both partners working long hours) is the most commonly adopted schedule for young nonparents. In contrast, the most common strategy for couples who have preschool-age children is to adopt either a neotraditional arrangement (the wife scaling back her work commitments) or for both partners to work less than long hours. This arrangement remains largely the same until children mature and leave the household (Moen and Sweet 2003). Most Americans espouse (at least in words) fairly egalitarian attitudes regarding women's and men's opportunities, believing that all members of society should have the same options and equitable rewards. But Exhibit 5.7 reveals gendered responses to time strains. When faced with a deficit of time available for care work, families tend to revert to the patterns laid out in the 19th and 20th centuries, normative arrangements that push women into the home and men into the workplace. The result is the perpetuation of gendered opportunity structures, arrangements in which men have careers while sacrificing family time, and women sacrifice careers for the rewards of family.

These time strains are resonating in a variety of other ways. Nearly half (43%) of all workers feel that they do not get enough sleep, and most workers do not feel highly successful at "balancing" work and family (Hochschild and Machung 1989; Moen, Waismel-Manor, and Sweet 2003).[5] Faced with the choice of "making a career" or "making a family," Americans are now having fewer children, having children later in life, or sometimes even giving up on family altogether (Altucher and Williams 2003; Farnesworth-Riche 2006). Consider the startling fact that women now have their first child nearly 4 years later than they did half a century ago. According to Current Population Survey data, in 1975, only one in 10 (10%) women age 40 to 44 had never had a child, but by 2008, this figure had risen to nearly one in five women (19%). Families that do have children are increasingly obliged to make use of child care providers (e.g., professional caregivers, neighbors, relatives), even though they are often reluctant to do so. Job demands appear to be having a profound impact both on the shape of family lives and on the prospects of forming families in the first place.

Some workers are demanding changes, requesting that their employers rework schedules to accommodate personal needs. Technological innovations and the rise of service work have made it possible for many jobs to be performed in settings outside the workplace (e.g., the home, in transit) and at a variety of times. These changes open the prospects for **flextime** and **flexplace** schedules that enable workers to labor at home, to negotiate the number of hours expected, to refashion the starting and ending times of work as well as break lengths, and to create opportunities to bank time.

Advocates of this approach to redesigning work point to the fact that it does not necessarily involve reducing how many hours jobs are expected to take, but does require expanding the amount of control workers have in setting schedules that match their needs off the job. As discussed in Chapter 2, the extent that these arrangements are available to workers remains limited and uneven. Some workers, particularly those with scarce skills, have been able to arrange reduced work schedules without suffering the severe economic and career penalties usually linked to part-time work (Meiksins and Whalley 2002; Tilly 1995). Although flexibility offers new opportunities to bridge work and family more effectively, new work arrangements also can create new concerns for working families, especially in respect to establishing boundaries that enable one to disengage from work. As some workers have found, the option of working into the evenings or at home creates new portals through which work can intrude into leisure time and domestic life (Golden 1998; Kossek and Ozeki 1999; Nippert-Eng 1996).

Conclusion

In the new economy, honor is commonly bestowed on those employees whose cars are the first ones in the parking lots and the last ones out. Those who cannot keep up with escalating job demands, or meet the higher requirements to get jobs in the first place, are forced to the margins into jobs that offer much more limited rewards and opportunities. For the growing number of salaried workers, there is no cap on the number of hours that employers can demand or limits on the expectations of what it takes to keep a job.

Even though organizational and technological innovations have opened up possibilities to liberate work (such as through the use of flexible arrangements), they are being implemented in a culture that still equates productivity with time spent at work and that fails to prevent overwork. As a result, requests for shorter hours, or failures to put in "face time," usually come with career penalties. As American culture has embraced overwork as a virtue, the new economy has unleashed new sets of pressures (largely unchecked) to labor longer hours with intensified effort. This in turn has helped to create work poverty, limiting the availability of jobs for others who want them.

For workers at the bottom of the opportunity structures in the new economy, schedules and wages intertwine in problematic ways. Current minimum wage standards, for example, do not require employers to structure work schedules in a manner that enables employees to make a living; the standards require only that employers pay a (woefully insufficient) base hourly wage. Even the more ambitious efforts to ensure a living wage (estimated at roughly $9 to $11 an hour in most locales) assume that workers who receive this wage would have the ability to work full time. Little public discussion has been directed to the need to provide adequate compensation and supports to those who need to work less. Many workers—such as single parents—cannot maintain full-time schedules and perform their family responsibilities outside the workplace. Moreover, living wages are, in almost every instance, calculated on the basis of day-to-day needs, leaving long-term needs such as paying for college tuition or saving money for retirement unaddressed. Nor are the problems of the many Americans who lack sufficient, steady work being addressed adequately.

Resolving these issues requires tackling—head on—the institutionalized practices that push workers to labor long hours and countering cultural values that keep Americans from putting work in its place. In some respects, contemporary workers are facing as profound a transformation in their society as their counterparts did during the Industrial Revolution. Rethinking the concept of a fair day's work—how much, how hard, and how long employees should be expected to labor—needs to be front and center in a national dialogue on ways to refashion the new economy.

Notes

1. Before becoming an academic, one of the authors of this book was a professional carpenter and worked for a company that required him to haul tools from the equipment shed and make them operational before he could "clock in," and wrap up equipment after he "clocked out." This added more than a half an hour of uncompensated labor to each day.

2. When the CPS calculates average work hours, it excludes those who are not in the labor force, which can create an impression of more time spent pursuing paid labor than actually occurs. The estimates reported in Exhibit 5.3 include working-age adults who are out of the labor force and who are categorized as working zero hours. This reduces the estimated number of hours worked, but enables us to reveal the remarkable changes in the average working hours created by women's entry into the paid labor force.

3. Long vacations, however, can create their own problems. Such was the case when a heat wave enveloped Europe in 2003. Because the French customarily take holidays during the summer months, elderly citizens had scarce access to health services, resulting in high mortality.

4. One in three motor vehicles in the world is located in the United States, and the U.S. vehicle-to-person ratio is nearly double that of countries in Western Europe (Freund and Martin 2000).

5. Authors' analysis of the Ecology of Careers Study. Data represent a sample of middle-class, dual-earner couples.

Chapter Six

Gender Chasms in the New Economy

Today, the number of women participating in the labor force approximately equals the number of men, and women's contribution to family finances is critical for household economic stability. Yet, gender remains a significant force in shaping the contours of opportunity in the new economy. It affects aspirations, shapes access to jobs, influences the compensation received, sways relations with coworkers, affects opportunities for promotion, and complicates labor force attachment. When viewed from a global perspective, gender has a greater impact on life chances than any other characteristic attributed to individuals (Fuchs-Epstein 2007). While *sex* and *gender* are commonly treated as synonymous, it is important to recognize that in the world of work, gender is not something that is simply possessed, it is something people do to each other. Gender creates the expectation that women and men will be capable of different things, and conflating gender with biological sex makes it seem "natural" to divide labor along gender lines. Throughout this chapter we call this assumption into question. Furthermore, we suggest that while gender chasms in labor markets and at work subordinate women in terms of both power and resources, they also mean that men lose out as well.

To illustrate these complex processes, in this chapter we consider why women and men gravitate to different types of occupations and how gender intersects with the capacity to perform work inside and outside of the home. We argue that the new economy has expanded and diversified employment opportunities for both women and men, but also that enduring forces continue to segregate workers on the basis of gender and impose hurdles that block mobility. Some of the chasms that separate men and women in the workplace can be attributed to the choices people make concerning what careers to pursue and where to concentrate their energy. But these gaps also are established by differential treatment on the job and by the various ways workers are subject to discrimination. Some inequalities are the product of interpersonal dynamics, but others result from job designs, as well as the value placed on different types of work.

While equity is a major reason to consider gender inequality, we also ask you to consider a related problem—how work in the paid labor force affects the capacity to provide care. In a society where most adults are expected to work outside the home, developing new means to support the young, the old, and the disabled has become a pressing policy concern. For example, recall in Chapter 1 the brief discussion of Eileen (the professional engineer), who had to make some difficult choices between her job and her family. Her experiences reflect those of the baby boom generation—and now some of their daughters—who entered the American workforce in unprecedented numbers in the latter part of the 20th century. On the one hand, Eileen's story is one of liberation, because she successfully entered into a career that had been largely sex-segregated in the old economy. On the other hand, many aspects of Eileen's life seem anything but liberated because work takes its toll on her ability to find the time to spend with her children and to have a life outside of her job. Ask also whether Eileen is a typical female worker in the new economy. Consider that Jamal's mother—a poor African American woman with drug dependency problems—matured with the same birth cohort as Eileen. Her different circumstances, and those of her son, illustrate how the gendered experiences of work in the new economy remain intertwined with the consequences of race and class. As we discuss below, these dynamics affect not only workers, but also their families and their communities.

When Did Home Work Become Nonwork?

Our friend Erika is a stay-at-home mother who left her job to support her husband's career and raise their two young children. While Kevin is at work in his research lab at a prestigious university—a job that requires long hours and frequent travel—Erika manages the home front—tending to scraped knees, arranging play dates, and doing all the other things that keep their family integrated into their community. For his efforts, Kevin receives unemployment insurance, retirement benefits, and a handsome salary. Although she receives occasional pats on the back from her friends who speculate about "how difficult it must be to stay home with the kids all day," Erika's social status is decidedly lower than that of her husband. She receives far less social recognition for her efforts, no awards, and no pay. Her status as a nonworker is most tellingly revealed when she is asked about her intentions to "go back to work" after her children get older. Her numerous frustrations with mothering work (which are more serious than her culture admits) are felt alone and in private. As a result, Erika is beset with ambivalent feelings about her career choices. Though taking pride in her work as a mother, she misses the rewards of having a good job in the world of paid work. And her husband also loses out, not in the workplace, but in having only brief windows of time to spend with his children.

Erika and Kevin have what is commonly considered to be a "traditional" household arrangement. In reality, when viewed in a historical perspective, this clear-cut separation of husbands' and wives' roles is anything but traditional. For most of human existence, almost all work centered in and around the home. In these **household economies**, there were gendered divisions of labor, with some tasks primarily assigned to women (cooking and child care being the most notable), but both genders contributed to the family economy. Shared responsibilities also were common. For example, in colonial America, when husbands became ill, wives served as "deputy husbands" and assumed responsibility for virtually all the activities previously performed by their spouses. Though women were not the political equals of men, they were considered workers and their efforts were considered absolutely essential to family survival (Boris and Lewis 2006; Boydston 1990; Ulrich 1982). Furthermore, the culture did not make a distinction between "going to work" and "going home," because work and family life were, to a great extent, one and the same. As a result, women who worked in and around the home were defined as being real workers, and their efforts were visible and socially recognized. A close approximation of agrarian work-family life worlds can still be witnessed on family farms, where the work of wives and husbands intertwines in proximate physical spaces (Cohen 1985).

Following the Industrial Revolution, and during the late 19th and early 20th centuries, American culture reconfigured its orientation toward home work and embraced a breadwinner-homemaker arrangement. But this did not happen immediately with industrialization. In fact, when the early factories physically separated paid employment from work around the home, the first workers were often women and children. Men sometimes refused to participate in these new arrangements (Hareven and Langenbach 1978). However, within the two succeeding generations, men's and women's roles became sharply differentiated. The new system, supported by the **ideology of separate spheres**, cast men in the role of wage earners, which in turn encouraged them to evaluate self-worth in terms of career success and the ability to provide for their family's economic needs. Many women, however, were expected to stay in the home and tend to the needs of their spouses and children. Although this husband/breadwinner–wife/homemaker arrangement is commonly termed *traditional,* it is actually a modern arrangement unique to the new industrial order. Its influence had major effects on divisions of work in the home, women's access to jobs in the paid economy, and social policy. Women were cast as the weaker sex, were considered dependents rather than workers, and were thought of as being less capable of "real work" than men. A landmark Supreme Court decision (Muller v. Oregon, 208 U.S. 412 [1908]), for example, ruled that it was not only acceptable, but also desirable, for employers to limit the number of hours their female employees worked. These are the words of Justice David Brewer, who wrote the majority opinion:

> That woman's physical structure and the performance of maternal function place her at a disadvantage in the struggle for subsistence is obvious . . . and, as healthy mothers are essential to vigorous offspring, the physical well-being of a woman becomes the object of public interest and care in order to preserve the strength and rigor of the race (Boris and Lewis 2006:81).

Justice Brewer's comments reflected the dominant cultural attitudes of the early 20th century that a woman's place is in the home. This ideology of separate spheres not only assigned home work to women and legitimated discrimination against them, it also redefined the economic value of women's efforts in the home. Rather than the home being a place of work (as it was in the 18th century and before), the home came to be viewed as a haven from work, the place where men recovered from the toils of the factory and the office (Lasch 1995; Zaretsky 1986). What was previously considered work was recast by an emergent **cult of domesticity** that asserted that household tasks were something other than labor, could be effortlessly performed, and offered so many intrinsic rewards that financial compensation was not necessary.

Even today, some cultural and political leaders call for a renewal of policy to support the supposedly natural arrangement of the husband/breadwinner–wife/homemaker model (e.g., Santorum 2006). Their vision of a traditional family corresponds with how families were represented on television shows that aired in the mid-20th century, such as *Leave It to Beaver, Ozzie and Harriet,* and *Father Knows Best.* These archetypes were of young, healthy, self-sufficient, heterosexual couples that could prosper on the efforts of one breadwinner. But as Betty Friedan documented in her pathbreaking book *The Feminine Mystique* (1963), home work in the 1950s bore little resemblance to the image presented on television. For most stay-at-home mothers, home work entailed a grinding repetition of alienating tasks, social isolation, and subservience. Idealizing these families reflects inaccurate ahistorical assumptions about the merits of the gendered division of labor and the organization of household work and the fact that the husband/breadwinner–wife/homemaker arrangement was largely restricted to two-parent families who had opportunities to get good jobs. This arrangement was not an option for those who were working in low-paying jobs, which included many immigrants and members of racial minorities (Coontz 1992, 1997; Gerstel and Sarkisian 2006). Class biases are evident as well, as middle-class men typically earned incomes that enabled them to support an unpaid wife, but working-class and poor men did not. Furthermore, the notion that women should be "free" from work to tend to matters in the home was not applied to all families, as many African American women were clearly expected to work as servants in support of the new ideal household arrangements for a privileged white society (Dill 1988; Glenn 2002). Early 20th century immigrant women were also far more likely than were their native-born counterparts to work outside the home. When they attempted to conform to the dominant view that women should be homemakers, they encountered financial hardship (because their husbands' incomes were often low). In response, many found ways to make significant economic contributions to the household *within* the home by taking in laundry, managing boarders, or by producing clothing and other necessities (Amott and Matthaei 1996).

In sum, the problems faced by Erika are a legacy of her culture's adaptation to industrialization, but are also specific to her racial identity and class position in the new economy. For women in less desirable economic circumstances, the prospects of choosing to stay home and mold their lives to conform to the mythic "traditional" arrangements were (and are) not commonly available. Since the 1950s, expectations regarding middle-class white women's paid employment have changed significantly and the cult of domesticity has weakened. However, the assumption that women will have primary responsibility for the management of the home remains. Moreover, this work is still not recognized for the vital economic role it plays in preparing the next generation for work and enabling workers (like Kevin) to put in long, undistracted hours on the job (Crittenden 2001; Kanter 1977).

Women's Participation in the Paid Labor Force in America

While the husband/breadwinner–wife/homemaker arrangement was only available for a portion of the families in the United States in the old economy, it was a dominant model for organizing family lives. Women, especially married women with children, were far less likely to work than were men. And, although this arrangement presented very different opportunities, the clear-cut gendered division of labor created a means to maximize collective family resources for the middle class (Becker 1981). Today, this arrangement is less common and women are almost as likely to work outside of the home as men and most married-couple households contain two working adults. Exhibit 6.1 shows the magnitude of this change. In 1940, only one in four women was in the paid labor force, but by 2010, nearly two in three were. Men's participation declined slightly, reflecting the aging of the population (there are now more retirees) as well as increases in the numbers of young people who delayed their entry into the labor force to complete school. Clearly a convergence is occurring with respect to the labor force participation of men and women.

Exhibit 6.2 shows that the normative arrangement of the old economy, in which the husbands went to work and the wives stayed home, is now the exception. In 2009, most married couples in the United States were dual earners, with both the husband and wife in the paid labor force. There are several reasons for the increase in women's labor force participation and the growing numbers of dual-earner couples. First is the changing perception of women's place in society and the remarkable transformation in gender role ideologies. But changes in the structure of the economy have been equally important. Stagnant male incomes made it increasingly difficult for the

Exhibit 6.1   Men's and Women's Labor Force Participation Rates (Age 16 Years and Older): United States 1940–2010

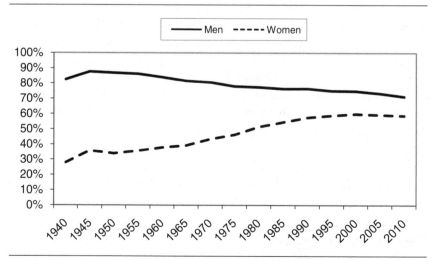

Source: Historical Abstracts of the United States, Bureau of Labor Statistics.

average husband to support a family on his wage alone (a situation long familiar to immigrant and minority families). Married women, whether they wanted to or not, found that they *had* to work if their families were to maintain their expected standard of living (Bernhardt et al. 2001; Gerson 2001; Warren and Tyagi 2003). Also, the remarkable growth of the service sector (a significant employer of female workers even in the old economy) created a demand for an expanded female labor force in jobs that do not challenge conventional attitudes about gender (Reskin and Roos 1991). Exhibit 6.2 also shows that married couples with children (including children under 6) are *more* likely than married couples generally to have two earners in the labor force, a remarkable reversal of the earlier preference for stay-at-home mothers. However, among married couples with only a single earner, it remains the case that wives are more likely than husbands to be the ones to drop out of the labor force; in that sense, children still affect the labor force attachment of wives more than husbands.

The rise of the dual-earner couple has forced families to confront issues that were less commonly present in the old economy, such as who will care for children while both partners are at work, whose career will take priority, and how to select and pay for day care. As we discuss later, old gender templates play a strong role in determining couples' responses to the stresses they experience in the new economy.

Exhibit 6.2   Employment Configurations of Married Couples: United States 2009

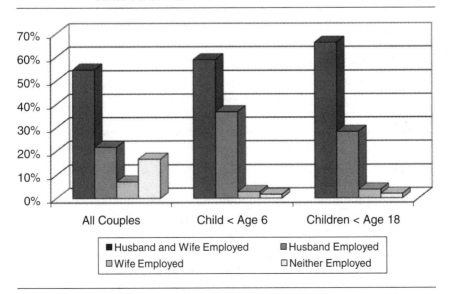

Source: Statistical Abstracts of the United States.

Gender Inequalities in Compensation

Men and women still face very different prospects of ever rising to the top of career ladders, falling off those ladders, or even making it beyond the lowest rungs. The data are sobering. On virtually every measure of earnings, women trail well behind men. In 2009, the average woman who worked full time earned $10,849 less than the average full-time male worker. Women are twice as likely to be poor and are much more likely to hold low-paying jobs that offer slim prospects for upward mobility (Gilbert 2003; Kalleberg, Reskin, and Hudson 2000; Padavic and Reskin 2002). They are half as likely as men to make it into the ranks of top income earners, and in 2009, for every three women chief executive officers (CEOs) of major corporations, there were 97 men (Branson 2010). Simply stated, while women are catching up to men in respect to labor force attachment, they have not caught up in respect to pay and position in the new economy. Recent trends indicate that the gaps are not narrowing as much as one might hope and some of the gender gaps remain as wide as ever.

Many complexities are involved in analyzing the extent of gender inequalities in earnings. For example, should one compare all women with all men, or only those who are working in the paid labor force? Note that the latter analysis tends to underplay the extent of gender inequality because women who are not in a position to earn any income are excluded from the calculations. The fact that women are much more likely to be working in part-time jobs (often because of their scripted gender roles) creates additional biases in wage disparity calculations that base analysis on full-time workers. But even when the comparisons of men and women workers are restricted to those who are working full-time, year-round jobs, women still tend to earn considerably less than their male counterparts, a dynamic revealed in Exhibit 6.3.

First the good news. Exhibit 6.3 shows that the women-to-men earnings gap has been closing. As the top line in the graph shows, in 1960, for every one dollar a man (working a full-time job) earned, women earned about 61 cents. In 2009, full-time women workers were earning 77 cents for every dollar earned by full-time working men. Now the bad news. A sizable wage gap still exists between men and women full-time workers. Again, because we have restricted the analysis to full-time workers, note that this analysis underplays the differences that would be revealed if part-time workers were included in the equation. In addition, while women's wages have been rising, much of the reason for the reduction in the gap between men and women workers is stagnation in men's wages. In other words, while women seem to be making some progress in enhancing their incomes, men's real incomes are actually lower than they were in the early 1970s. So the closing

Exhibit 6.3   Women's and Men's Earnings and Income Ratios for Full-Time
Year-Round Workers: United States 1960–2009 (Earnings Adjusted
to 2009 Dollars)

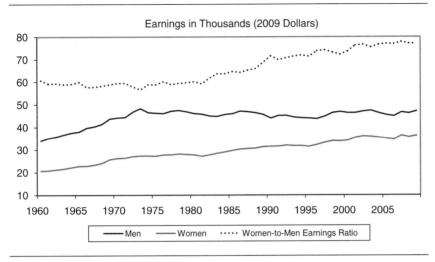

Source: U.S. Census Bureau (2011).

wage gap can be explained both by an enhancement in women's earnings, and a leveling of men's earnings. Also, the most recent decade of data suggests the rate of convergence between men's and women's earnings is slowing, at least in comparison with the gains made from 1970 through 1995. The reasons for this slowing convergence are not altogether clear, but one study suggests that it may be the result of contemporary women pursuing jobs that offer lower pay (Blau and Kahn 2006), a dynamic discussed later in this chapter.

Of late, a number of mainstream media publications have considered an interesting trend, as approximately one in five wives make more than their husbands, a remarkable change considering that in 1970 it would have been only one in 20 couples in a comparable situation (e.g., Perpitone 2010). However, more careful study of this dynamic indicates that the change is not as dramatic as these figures suggest. According to research by Sarah Winslow-Bowe (2006), if one were to follow couples in which the wives outearn the husbands over time, within a few years the vast majority shift to the husband having a higher income. The reason for this is because the women's income advantage commonly occurs because the husband has lost his job or is temporarily out of the labor force. And if one were to look at those couples in which the wife has a consistent income advantage, most of these couples are low-earners. While the wife might make a significantly higher percentage of the household earnings than the husband in these couples, the actual monetary difference in their incomes is quite small, given that both are earning low incomes. It is actually quite rare to find couples in which the wife maintains an income that is consistently higher than her husband's over time.

In sum, wide gaps exist between men's and women's earnings. Some of the explanation can be found in the fact that household work is not compensated, which leaves many women without paychecks. The expectation that women will assume disproportionate household and child care responsibilities increases the odds that women will be funneled into part-time jobs, which also results in lower earnings. But even among full-time workers, women's pay lags well behind that of men. Why is this?

Socialization, Career Selection, and Career Paths

The large-scale entry of women into the paid labor force has reshaped how men and women view their roles and capabilities, as well as opened doors that were previously closed. As a result, the gender composition of entire fields has shifted. For example, women are now the majority of graduates in veterinary colleges, a profession that in the 1960s and before was exclusively male. And men are now more apt to express a strong desire to spend time at home with children and are beginning to enter fields (such as nursing) where they were previously nearly entirely absent. Nevertheless, gender segregation remains a central part of the contours of work in the present-day American economy. One study estimated that half of all American working women would have to change fields to eliminate the disparities in the gender composition of occupations (Jacobs 1999).

One reason for these imbalances is that gender templates encourage men and women to have different aspirations. From early childhood, individuals make choices about what kinds of activities to engage in, what skills to develop, and what interests to pursue. Along each step of the way, they develop skills that, in turn, become part of the toolkit they use to construct future encounters. Ultimately, these choices influence eventual occupational goals and destinations. These choices are made in the context of powerful social expectations about gender. Various **agents of socialization** (including parents, schools, and the media) instill beliefs that boys/men and girls/women are not equally suited to all tasks. The confluence of social pressures and the transmission of taken-for-granted paths, over time, explains why men and women place different emphases on their identities as breadwinners, why they select different occupations, and their willingness to sacrifice their careers for the needs of their children and spouses. The result of socialization is that boys and girls (and men and women) are continuously molded to have interests and identities suited to different types of endeavors (Bem 1993; Chodorow 1978; Lorber 1994; Lott and Maluso 1993).

Americans are socialized to expect women to fill certain positions in society, and in that respect many jobs are "gendered." Not every culture views these lines of work in the same way. For example, in the United States most secondary school teachers are women (60%), but in India, two in three (69%) secondary school teachers are men (United Nations Statistics Division 2011). Similarly, three in four doctors in the United States are men, but in Russia nearly three in four doctors are women. Statistics such as these firmly establish that expectations about which jobs should be done by men or women are culturally determined (Harden 2001). Although women have made remarkable forays into many professions that were previously nearly exclusively male, Exhibit 6.4 shows that many occupations remain almost exclusively filled by women workers. Notably, nearly all preschool and kindergarten teachers are women, as are secretaries, dental hygienists, dental assistants, dietitians, typists, and child care workers. Conversely, women are nearly entirely absent in construction industry jobs such as those performed by plumbers, carpenters, and electricians.

Gender expectations interact with class and racial differences to produce a complex, varied pattern of gender segregation in American workplaces. For women at the bottom of the opportunity divide, a major problem is the existence of **occupational ghettos**, gendered jobs that typically offer few paths to upward mobility (Grusky and Charles 2004). As Exhibit 6.4 shows, most maids, day care workers, and secretaries are women. The fact that the income gap between white and black women is relatively small (far smaller than it is for white and black men) is linked to the fact that women of all races are funneled into lower-paying jobs such as these.

Socialization encourages workers to feel well-matched to gendered lines of work. For example, one study of secretaries found that most of these women (especially older women) enjoyed their jobs. Rather than perceiving gender ghettoization as a source of oppression, they had accepted what they were socialized to believe—that men and women are different from one another, that women are skilled at doing tasks that men are bad at performing (especially nurturing), and that they themselves possessed these skills (Kennelly, 2002, 2006).

Men with lower levels of education traditionally enjoyed advantages over women with comparable credentials in the old economy. These men were encouraged to seek, and generally found, jobs in manufacturing, construction, or automobile repair—work that offered opportunities for skill and income expansion and stronger prospects for economic security

Exhibit 6.4   Occupations With the Highest Percentages of Women Workers: United States 2009

Occupation	Percent Female	Number Employed (1000s)
Preschool and kindergarten teachers	98	691
Dental assistants	98	286
Secretaries and administrative assistants	97	3176
Dental hygienists	97	137
Speech-language pathologists	96	140
Child care workers	95	1292
Bookkeeping, accounting, and auditing clerks	92	1306
Word processors and typists	92	163
Registered nurses	92	2839
Teacher assistants	92	1006
Receptionists and information clerks	92	1277
Licensed practical and licensed vocational nurses	91	633
Medical assistants and other health care support occupations	91	776
Hairdressers, hairstylists, and cosmetologists	90	800
Billing and posting clerks and machine operators	90	481
Maids and housekeeping cleaners	90	1428
Health care support occupations	89	3309
Medical records and health information technicians	89	103
Nursing, psychiatric, and home health aides	88	2002
Dietitians and nutritionists	87	99
Occupational therapists	87	106
Tellers	87	432
Information and record clerks, all other	87	109
Payroll and timekeeping clerks	87	150
Eligibility interviewers, government programs	87	63
Hosts and hostesses, restaurant, lounge, and coffee shop	86	258
Special education teachers	86	385
Paralegals and legal assistants	86	354
Massage therapists	86	150
Personal and home care aides	85	926

Source: U.S. Bureau of Labor Statistics (2010).

in both the short and long term. But less educated African American men were a notable exception because they often were excluded from jobs of this kind. In recent years, the perception has grown that the situation of boys from poor and working-class backgrounds has deteriorated and that girls now have the advantage. The decline of well-paid manufacturing jobs, and evidence that girls are outperforming boys in school, seem to indicate that boys who do not have high levels of education are aspiring to jobs that no longer exist. However, a careful look at the evidence indicates that race, rather than class, is interacting with (not replacing) gendered aspirations. Less-educated white men are not aspiring in large numbers to enter traditionally female fields but continue to find desirable, "masculine" work in areas such as construction or criminal justice. Poorly educated African American men (like Jamal, the young worker introduced in Chapter 1) continue to hope for traditionally male jobs that are unavailable to them and thus are falling behind (Kimmel 2006; Young 2003). Gender segregation at the bottom of the opportunity divide persists, in other words, but takes different forms for members of different racial groups.

Gender socialization also plays a role for men and women holding comparatively good jobs in the new economy. One means of illustrating its power is to consider how young adults select college majors, a critical step in determining subsequent career options. Exhibit 6.5 reveals that even today, young men continue to gravitate toward traditionally male-dominated fields (e.g., computer science and engineering) and women tend to gravitate toward college majors associated with helping professions (psychology, education, health care). In part, this is the result of gender beliefs, one of which is that men tend to be better at math than women. Research into this issue reveals that differences in mathematical abilities are actually small, or even nonexistent. And even when differences are identified, scant evidence exists to support any conclusion that these are based on biology (Ceci, Williams, and Barnett 2009). But these beliefs create self-fulfilling prophecies. Once the perception of capabilities is embraced by young men and women, even when those individuals have equivalent skills, it influences their aspirations to pursue mathematics-related professions. In other words, boys pursue math-related careers not because they actually have better skills than girls; they do so because they *believe* that they are better (Correll 2001, 2004). These beliefs, in turn, lead them to take additional courses and further develop their skills. But this is not the only factor in play. As we discuss later, those women who do enter male-dominated fields (such as engineering and science) are commonly marginalized, or find that these jobs are incompatible with their family responsibilities (Ceci, Williams, and Barnett 2009; Committee on Maximizing the Potential of Women in Academic Science and Engineering 2006; Preston 2004).

Exhibit 6.5   Gender Compositions of Bachelor's Degrees Conferred: United States 2007–2008

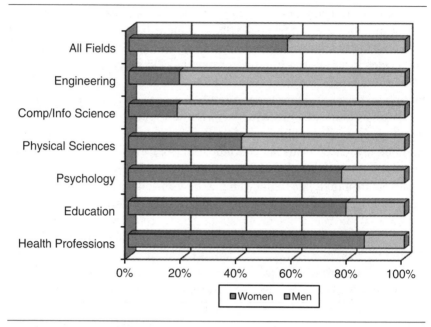

Source: Chronicle of Higher Education Almanac, 2010 edition.

Aside from its role in career selection, gender socialization influences how men and women respond to work and family strains. Among working-class and poor families, women are more likely than men to arrange work schedules to allow involvement in their children's lives (Garey 1999). Among higher-income families, men seldom sacrifice their careers for family (Becker and Moen 1999). In fact, the opposite is the case, as men devote themselves more intensively to their work because they see this as a duty to their families. To compensate for their husbands' reluctance or unwilling-ness to relinquish career goals, women modify their work lives, scale back their work hours, and redirect their professional interests toward alternate careers (Moen and Sweet 2003; Wharton 2002). As a result, dual-earner families are transforming the 20th century separate spheres model into new **neotraditional arrangements,** wherein women retain primary responsibility for child care while keeping one foot in the labor market in lower-pressure (and typically lower-paying) jobs.

The strategic choice to scale back work hours comes with significant costs to professional women, not only in immediate compensation, but also in long-term career prospects. Working women who are able to arrange shorter hours or flexible schedules with their employers are often placed on **mommy tracks** and are assigned tasks that offer fewer rewards and less opportunity for growth compared with workers on the fast track (Barnett and Gareis 2000). And even when more attractive opportunities to scale back on work are present (which can be the case for professional workers such as lawyers and academics), there is a very real prospect that they may never get back on track once they step off (Meiksins and Whalley 2002; Moen and Roehling 2005).

For dual-earner couples, neotraditional arrangements create marked disparities between wives' career opportunities and those afforded their husbands. As one partner's job is put on hold, the other's continues to grow, which reverberates into subsequent life-course decisions, such as whose job to favor, when and where to move, and who should assume responsibility for family care work. Exhibit 6.6 illustrates this dynamic. When dual-earner couples manage their relationships, they can choose to favor either one partner's career (leaving the other as a trailing spouse), to take turns, or to give priority to neither partner's career. Most husbands and wives start their careers on relatively equal footing, with neither partner's career being favored, but as their lives develop over time, wives' careers tend to be towed along and follow the direction charted by their husbands' job opportunities. Their options are constrained by investments in husbands' careers, which in turn limits their options to advance and grow, even after children have left the household (Pixley 2008; Sweet, Moen, and Meiksins 2007).

Exhibit 6.6   Percentages of Husbands and Wives Reporting That Their Career Was Favored Over Their Spouse's Career (by Life Stage): Dual-Earner Professional Couples

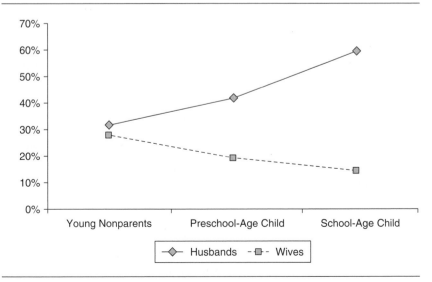

Source: Sweet and Moen (2006).

Catherine Hakim (2001) argued that the reason for this type of dynamic is that many women express a preference to center their identities in the home rather than in the workplace and that this may explain some (but not all) of the divergences in husbands' and wives' careers (Kan 2007). It is important to recognize, however, that for most women the decision to leave the labor force, or to scale back on career aspirations, is not a happy choice, but one that is necessary when the combined demands of work and family outstrip time and energy resources (Stone 1997). When faced with these types of tensions, a common strategy is to give up either on career goals or family goals, and thus it is not so much that women prefer to drop out of the labor force, but that this is the preference among very constrained options. Alternatively, one can redefine these goals or locate work situations that are more compatible with responsibilities held outside of the job. For example, some women use self-employment as a means to balance work and family commitments, and these types of jobs are commonly pursued after the birth of children. Such is the case for African American female entrepreneurs who establish beauty salons (Harvey 2005). In contrast, children have little effect on men's pursuit of self-employment, and men tend to use self-employment as a means to advance career objectives (Carr 1996). Interestingly, a more recent study revealed that a woman's social class may affect her reasons for self-employment. The study found that women in lower-level jobs pursue self-employment as a means of balancing work and family, but women pursuing higher-powered professional careers act like men, and when they do become self-employed, do so primarily for career reasons (Budig 2006). Preferences may be changing, however. In interviews with young women about their career aspirations, Kathleen Gerson (2009) asked young women what they wanted out of life. Not surprisingly, most wanted a family (including children and a spouse), as well as a career. When she pressed them and asked, "suppose you could only have two of these things . . . a husband/partner, a child, or a career, which two would you pick?" most of the young women chose to discard the husband!

In sum, socialization continues to shape men's and women's career pursuits, as well as their expectations about what they should do off the job. The tensions evident in the new economy tend to push workers to revert to the gender templates established in the old economy, with women assuming primary responsibility for the management of the domestic sphere, and men adopting roles as breadwinners. But this adoption of old gendered strategies for managing work and home occurs in a new economic and cultural context in which most workers want and need to remain attached to the paid labor force. Liberating work in the new economy will require a serious response to the culturally based forces that push working families to adopt neotraditional arrangements—a practice that tends to cost both men and women the opportunity to work as equals inside and outside the home (Deutsch 1999).

Interpersonal Discrimination in the Workplace

Socialization contributes to gender inequalities at work by influencing the supplies of men and women workers seeking entry into different fields. But gender also plays a role in shaping the demand for workers with specific qualities. To illustrate how this happens, imagine interviewing a series of applicants for a demanding job, a position that will require long hours of work and some travel. Your choice comes down to two candidates, Jane and John, individuals who have identical qualifications and who performed equally well at all stages of the interview process. The only difference between the candidates is that Jane is visibly pregnant, but John (obviously) is not. Which person would you hire? We have posed this question to hundreds of students over the years, but only rarely has a student offered even mild support for hiring Jane. They have eloquently (and often emphatically) argued that her candidacy poses a number of concerns, including the inevitability that she will want to take time off from her job, that she will be unable to put in as long hours as John, and that she will be in no position to travel once her child is born. As uncomfortable as it may make them, they conclude that it would be their responsibility to hire the person most fit to meet the demands of the job—John.

All of these conclusions are based on **gender schemas**, stereotyped assumptions about women's behaviors and desires (Valian 1998). Notice how parental status becomes a master status for Jane in a way that it does not for John (who conceivably could have an infant waiting for him at home). There is a ready acceptance of the assumption that Jane would want to have reduced hours after she had her child and that the child would detract from her work. But it is also possible that she could have a partner who wants to stay home or has an excellent day care arrangement. What takes precedence in favoring John is the emphasis on crudely constructed visions of how women and men behave in different situations, and the assumption that Jane and John would behave in a corresponding manner. This is a process termed **statistical discrimination**, wherein perceptions of group tendencies become the rationale for differential treatment of individuals. Statistical discrimination is problematic for a number of reasons, not least of which is the widespread use of **stereotypes**—often erroneous assumptions about social groups that obscure the range of behaviors and abilities present within any population. An important study by Shelley

Exhibit 6.7   Instead of Hiring This Woman, Most College Students Would Hire an Equally Qualified Man. Would You?

Source: Copyright © by Getty Images. Reprinted with permission.

Correll (2007) found that when presented with comparable job candidates, evaluators perceived women as being largely as capable and committed as men. . . . that is, unless they are mothers. In contrast to the motherhood penalty, observed in respect to perceived deficits in competence and commitment, fatherhood status actually improved the evaluations of men. Likewise, when actual identical job applications were sent out, varying only in respect to parental status and gender, twice as many childless women were called back compared to mothers.

Women are sometimes subjected to **hostile sexism**, the belief that they are inferior to men at specific tasks (Masser and Abrams 2004). An example can be found in the comments made by former Harvard University President Lawrence Summers, who in 2005 expressed his opinion that one reason why female scientists are scarce at top research universities is because of innate differences between the sexes. When a power holder such as this accepts these beliefs, the likelihood increases that he will operate on the assumption that women are likely to fail and to view their accomplishments with skepticism (Valian 1998). For example, one study found that female scientists in the 1990s needed to be twice as productive as their "more promising" male counterparts to receive prestigious postdoctoral training opportunities (Wenneras and Wold 1997). The denial of the opportunity to succeed contributes to self-fulfilling prophecies, as workplace stratification systems lopsidedly allocate opportunities to men and then reward their successes with even more opportunities to excel.

Note that women also can be the objects of **benevolent sexism**, which operates on the assumption that they are better than men at other types of activities, such as planning social events, organizing files, or caring for children. This, in turn, prompts gatekeepers to open doors for their entry into jobs that have been traditionally defined as "women's work." The problem here is that women tend to be viewed as exceptionally qualified for work that is less rewarding than "men's work," an issue we return to shortly. Benevolent sexism also can lead employers or managers to "protect" female workers, and lead them not to challenge them or involve them in difficult or dangerous tasks. The result is that women may not have as many opportunities as men to develop new skills and to demonstrate their capabilities on the job.

Overt discrimination, job discrimination on the basis of gender, was a prominent contour of the old economy. This involved the visible, conscious, and intentional decision to bar women's entry into occupations. Until the mid-20th century, women were excluded from most professional occupational fields and it was customary for them to be paid lower wages than men, even when they performed the same work. Not until the late 1960s and early 1970s did elite colleges such as Dartmouth and Princeton admit women into many of their professional programs. Similarly, overt discrimination excluded women from well-paid jobs in the skilled trades. Entry into apprenticeship programs was difficult or impossible for women; union membership was largely restricted to men, and women who did succeed in entering male-dominated trades were often met with a hostile reception that made it difficult for them to "learn the ropes." Today, there are far fewer instances of overt discrimination, largely because of the **Civil Rights Act of 1964**, which prohibited employment discrimination based on sex (as well as race, color, religion, or national origin), and the **Equal Pay Act of 1963**, which prohibited paying men and women different wages for equal work. These laws, and a series of successful lawsuits against employers, helped to change work policy and employment practices, which in turn has opened avenues for women to enter historically male-dominated professions and industries. Overt discrimination still occurs in the new economy, but at least in comparison with the 1950s and before, the problem is less severe and women now have legal recourse.

When faced with overt discrimination, workers who know that they are being treated in an unfair manner are ostensibly in a position to level complaints or to sue their employers. Such was the case for Ramona Scott, one of 115 women who testified in a class-action gender discrimination lawsuit against Walmart. Scott testified that, after being passed over for promotion, her manager told her, "Men are here to make a career and women aren't . . . retail [work] is for housewives who just need to earn extra money." Stephanie Odle, an assistant manager, joined the same class-action lawsuit against Walmart only after she learned by chance that her male coworker earned $10,000 more a year than she did (Head 2004). In the summer of 2011, the U.S. Supreme Court ruled that such large class-action suits were unacceptable. A major concern for the court was the question of how reasonable it is to make awards to most or all women who work for an employer, based on a limited number of identifiable instances of problems in employment practices. The decision is likely to make it much more difficult to use class-action suits in cases of suspected employment discrimination affecting large numbers of employees (Liptak 2011). An added barrier in establishing class interests is that employees often do not know when and how they have been discriminated against, in part because information on employee pay is usually kept closely guarded. As a result, many workers never learn the extent to which they have been treated unfairly. Even female scientists at MIT had little understanding of the extent to which they were disadvantaged in work assignments, promotions, and allocation of lab space until a group of them shared their experiences and saw patterns that otherwise were invisible (Head 2004; Massachusetts Institute of Technology 1999; Myrdal [1954] 1995). Leveling complaints also can involve career costs because even when these complaints are legitimate, the victims can be perceived as troublemakers (Gallant and Cross 1993).

Even when they want to, employees face uphill battles in lodging complaints against their employers. One of the most notable cases in recent history is that of Lilly Ledbetter, a manager at Goodyear who repeatedly received lower raises than the men who worked with her. Over time, the accumulation of smaller pay increases led to her making tens of thousands of dollars less than men who were hired at the same time as she was, and who were working in comparable positions. In a stunning decision, the U.S. Supreme Court ruled (by a narrow majority) that Ledbetter was *not* entitled to damages under Title VII of the Civil Rights Act of 1964 because she did not protest the compensation differences shortly after each of the pay setting decisions was made. The fact that she lacked the information needed to do so, as well as the implications that seeking redress might have had on her career, were captured in the dissenting statement by Justice Ginsberg:

> The Court's insistence on immediate contest overlooks common characteristics of pay discrimination. Pay disparities often occur, as they did in Ledbetter's case, in small increments; cause to suspect that discrimination is at work develops only over time. Comparative pay information, moreover, is often hidden from the employee's view. Employers may keep under wraps the pay differentials maintained among supervisors, no less the reasons for those differentials. Small initial discrepancies may not be seen as meet for a federal case, particularly when the employee, trying to succeed in a nontraditional environment, is averse to making waves. *(See* Ledbetter v. Goodyear Tire & Rubber Co., *550 U.S. 618 [2007])*

To address this concern, the first act of legislation signed into law by President Obama was the Lilly Ledbetter Fair Pay Act, which changed how the statute of limitations for filing grievances against employers is defined. Rather than requiring that the grievance be filed within 180 days of the employer's decision in respect to *the determination* of pay levels, employees must now file a grievance within 180 days of *the receipt* of each pay check in which discrimination is argued to be evident. While a significant step forward, the onus remains on employees to show that they have been discriminated against and, as we discuss below, more often than not employees remain in a position that limits their ability to do so.

Another concern is that women and members of minority groups also can be subject to **covert discrimination**, wherein power holders structure opportunities in a biased manner, but in ways that the employee may find difficult to detect or prove (Benokraitis 1997). A manager who is reluctant to employ or promote women, for instance, can try to increase the chances of failure by making the employee's life exceedingly difficult. An employee may find herself short-staffed, her hours cut, or she may be assigned unfavorable tasks or scheduled for work at times incompatible with her needs off the job. Similarly, she may be set up for failure by being placed in situations for which she has not been properly trained or on jobs that have low odds for success (Zimmer 1986). Covert discrimination is one of the most frequently cited contributors to the existence of a **glass ceiling**—an invisible barrier that prevents female white-collar workers from rising to the highest ranks.

Discrimination is less often practiced deliberately *against* employees, but rather *for* employees whom managers see as especially worthy of opportunity. Among managers, "fitting in" is of paramount importance, which influences their tendency to gravitate to, and create, gender and racially homogenous work teams (Jackall 1989; Kanter 1977). White men hold

most positions of power, so they have the greatest leverage in reproduc-
ing gender and racially homogeneous managerial teams (Elliott and Smith
2004). The protégé, the junior employee who looks, acts, and identifies
with the (white male) boss, has a decided advantage in this culture and may
be rewarded by supervisors' designing new jobs fitted to his strengths. The
women excluded from "old boys' clubs" sometimes suspect that they have
been wrongly passed over, but proving the existence of discrimination in
such a culturally mediated process can be challenging (if not impossible).

An additional problem confronting women (as well as racial and sexual
minorities) can be the existence of a **hostile work environment,** a situational
context so caustic that it undermines the ability to perform jobs. Although
not formally barred from working in the company of men, women can find
themselves in uncomfortable situations or subjected to unwanted sexual
propositions. Fully one in three women in their mid-20s has experienced
unwanted touching or invasions of their personal space on the job (Uggen
and Blackstone 2004). Work in such contexts can send subtle and not-so-
subtle messages to women that they either do not belong or that they are
primarily valued as sexual property and conquests; it also can undermine
their productivity and advancement (Benokraitis 1997; Gallant and Cross
1993). Interestingly, similar reports of hostile work environments are
offered by men who try to enter traditionally female-dominated professions
such as clerical work or nursing (Henson and Rogers 2001). However, the
uncomfortable gender environment men encounter in these situations also
can produce **glass escalators** that lead them into higher-level, better-paid
jobs. Male kindergarten teachers, for example, may encounter a difficult
work environment (especially from hostile parents) in which questions are
raised about their motives, their sexuality, and so forth. Far from harming
their career prospects, this can actually encourage the transformation of
these teachers into school principals, a more conventionally male role in
which they acquire greater authority and higher pay (Williams 1991).

Some women are placed by their superiors in no-win situations, forced to
choose whether to acquiesce to unwanted relationships, or resist and experience near-inevitable, career-damaging repercussions (MacKinnon 1998).
Male-dominated workplaces tend toward masculine cultural values and
behavior, and in this world, sexual jokes and horseplay can be especially
problematic if only a few women are represented. This also poses problems
for sexual minorities, whose participation in sexual discussions or office
romance is unwelcomed and received with hostility (Giuffre and Williams
1994; Woods and Lucas 1993). However, policing sexuality in workplaces
of the new economy is problematic for a number of reasons. One concern
is that many consensual romantic encounters (and marriages) are formed
as the result of working together, and it is not unusual for spouses to seek

work within the same organization (Astin and Milem 1997; Creamer and Associates 2001; Schiebinger, Henderson, and Gilmartin 2008; Sweet and Moen 2004). Additionally, work cultures (accepted by both male and female employees), such as those of restaurant waitstaff, are often accepting of sexually laced coworker interplay, including pinching, joking, and even casual encounters (Williams, Giuffre, and Dellinger 1999). Other jobs, such as those of comedy writers and those in the publishing industry, require employees to develop materials that some may find objectionable, and this in turn may influence the gender composition of fields (Exhibit 6.8). Attempting to constrain discussions or eliminate potentially offensive materials could ultimately undermine the prospect of actually doing this work (Dellinger and Williams 2002). In the service economy, selling sexual titillation (such as at a Hooters restaurant) is part and parcel of some jobs.

How do most workers respond when they feel they are being discriminated against or find themselves in hostile work environments? Among those in low-level jobs, such as employees in the fast-food industry or retail sector, the most common response is to tolerate the situation, and when that fails, to try to find work elsewhere (Tucker 1993). But for women in higher-level positions, harassment or inequitable treatment on the job can confront them with a classic catch-22. Even in situations where the worker can document unfair treatment, registering a complaint (which typically would require complaining to someone in authority) tends to be taken as evidence of a failure to "fit in" or "get along" (Gallant and Cross 1993).

Exhibit 6.8   Daily Show Writers and Producers Receive Their Emmys in 2009: Why Are There So Few Women?

Source: Copyright © by Mario Anzuoni/Reuters/Corbis. Reprinted with permission.

Ultimately the boundaries of acceptable conduct in the new economy will be played out in the courts, as well as in the ongoing negotiation of informally established workplace cultures.

Structural Dimensions of Gender Discrimination

Thus far, we have considered the problem of gender discrimination largely by focusing on how individuals respond to one another on an interpersonal level. If gender inequality were simply a matter of interpersonal discrimination, solutions would involve changing people through initiatives such as training seminars or leveling fines against those who act irresponsibly. However, even if these initiatives were successful, deeper structural forces that reinforce gender inequities in the new economy would remain. In this section, we focus on two institutionalized practices that limit women's ability to compete in the modern workplace: the ways in which women's work and men's work are valued, and the ways in which job designs conflict with gender scripts.

The Devaluation of Women's Work

Consider the differences between two low-status occupations, one of which is primarily occupied by men, the other by women. The job descriptions in Exhibit 6.9 are quoted from the Bureau of Labor Statistics *Occupational Outlook Handbook, 2010–2011* edition, a handy source of information about the nature of different jobs and the economic returns workers receive for their efforts.

Why does a garbage collector earn $5.81 per hour more than a day care worker? One could suggest a variety of reasons, but none of them eliminates the fact that day care work is more challenging and requires greater skill. Even the argument that garbage collectors need greater pay to compensate for the dirty nature of their work crumbles when one remembers that providing care for young children involves more than a little dirt and quite a few unpleasant smells! Disparities such as these have led feminist scholars to consider the **comparable worth** of different types of jobs. Their research indicates that wage disparities are not simply the result of labor supply and demand; instead they are intrinsically tied to the valuation of men's and women's work. Of course, one should use caution in generalizing from a selective comparison of two specific occupations. To address this concern, Paula England and her colleagues developed a variety of statistical approaches to document how comparable worth affects women's

Exhibit 6.9   Two Different Job Descriptions: Why Does One Job Pay Less Than the Other?

Refuse and Recyclable Material Collectors
Gather refuse and recyclables from homes and businesses into a truck for transport to a dump, landfill, or recycling center. They lift and empty garbage cans or recycling bins by hand or operate a hydraulic lift truck that picks up and empties dumpsters. They work along scheduled routes. Median pay in 2008 was $14.93 an hour.

Child Care Workers
Nurture, teach, and care for children who have not yet entered kindergarten. They also supervise older children before and after school. These workers play an important role in children's development by caring for them when their parents are at work or are away for other reasons, or when the parents place their children in care to help them socialize with other children their age. In addition to attending to children's health, safety, and nutrition, child care workers organize activities and implement curricula that stimulate children's physical, emotional, intellectual, and social growth. They help children explore individual interests, develop talents and independence, build self-esteem, learn how to get along with others, and prepare for more formal schooling. Median pay in 2008 was $9.12 an hour.

Source: Bureau of Labor Statistics (2009).

earnings, considering not so much who is working but, rather, the jobs that are associated with men's and women's roles in society. England's statistical models show that if a man with the same level of education moved from an occupation that was entirely composed of men to an occupation that was entirely composed of women, his earnings would substantially decline. Jobs that require care work (which is culturally associated with women's responsibilities) pay 5% to 10% less than do jobs that do not, even if other relevant factors are considered (e.g., jobholders' education, years on the job, supervisory responsibilities). It is hard not to conclude that "women's work" is systematically undervalued in the paid economy (Aman and England 1997; England, Budig, and Folbre 2002; Karlin, England, and Richardson 2002).

There are often marked disparities in the comparable worth of *tasks* within jobs as well. Most occupations require workers to perform a wide range of activities, all of which are central to organizational success. New employees need to be trained, notes need to be taken, interpersonal conflicts need to be resolved, and a variety of other personnel and technological concerns need to be addressed on a day-to-day basis. Men and women who are employed in the same type of job may be directed to perform different aspects of that work. For example, women professors tend to be assigned a disproportionate share of student advisees, which in turn frees their male colleagues to spend more time on more highly valued research activities. Although colleges and universities give lip service to the importance of advising students (and cannot succeed without this work), the time spent on this work can hurt, rather than enhance, an employee's performance review. In the end, the tasks that men tend to be assigned carry greater weight in decisions about promotions and pay increases (Fletcher 2001).

How Job Designs Discriminate

Nearly all police departments require patrol officer candidates to demonstrate physical aptitude for the job and rely on tests that commonly include measures of physical prowess. On the surface, tests of agility and strength appear to be gender-neutral because the same standards can be applied to all candidates. But when one looks at the effects, and how fitness exams tend systematically to screen out women applicants, structural biases in their construction and application are revealed. At one police department in Pennsylvania, for example, only about one in 10 (12%) female applicants successfully passed the running test, compared with more than half (60%) of the men (Brooks 2001). On measures of strength, women also fare poorly. The average woman can only perform sit-ups and push-ups at about 75% the capacity of men, and the average grip strength for women is only 57% of that of men (Shephard and Bonneau 2002). These differences are real, but one must ask how often a police officer actually will need to perform push-ups or run long distances in the line of duty. One also wonders how much grip strength it actually takes to fire a pistol, and why tests of strength, such as these, are on fitness exams. Nevertheless, these are the standards commonly used to screen applicants.

Part of a patrol officer's job requires physical work, and applicants should not be exempt from those responsibilities. However, it should be the case that the tests used to assess applicants are *relevant* to the job. Most day-to-day police work involves mundane rides in patrol cars, writing up reports, processing people, and handing out tickets. When force is needed, it is commonly performed as a team effort rather than being performed by lone officers. Technologies such as Tasers also make the successful management of physical altercations depend less on strength. Perhaps most devastating to the argument regarding the need for physical tests for police work is the observation that physical tests are used to determine who can *enter* into police jobs, but not who can *keep* them.

Beyond these observations, consider how the existing selective testing techniques tend to ignore the skills that women are especially likely to possess. For example, only about half of all police departments include situational tests, wherein a simulated real-life encounter is constructed and recruits are judged on their ability to handle routine traffic stops or more dangerous encounters (Shephard and Bonneau 2002). In these simulation exams, the ability to read emotional states and intentions is the needed skill, a strength more commonly associated with feminine qualities than masculine attributes (Martin 1982). Arguably, interpersonal tests are more important than physical tests because the ability to read a person's motivations and emotional state, and respond accordingly, could help officers avoid having to use physical force in the first place. Because these tests are not as widely used, the likelihood that candidates with poor interpersonal skills will make it through the screening process is increased, which favors the candidacy of underqualified male applicants. Also, note that the work of police supervisors or detectives almost never requires the use of force, but the career path to those positions is through patrol jobs, which in turn requires demonstration of physical prowess.

Structural forms of discrimination are built into the very design of work, and therein lies the problem, because most jobs and professional expectations have been designed with men in mind. For example, to become a pilot in the U.S. Air Force, one needs to be 64 to 77 inches tall, a standard that is squarely in the midrange for American men, but that screens half of all women from eligibility.[1] Here we observe something different than the police exams, in that height criteria are not arbitrary or irrationally selected. Pilots need to be able to reach the levers and buttons that surround them on all sides of their cockpit seats. In this case, the structural discrimination occurred in the design of the plane itself, which was tailored to fit the typical man, not a typical woman.

The legal scholar Joan Williams (2000) identified a number of interesting cases such as these in her book *Unbending Gender,* which examined the templates used to construct job expectations. She concluded that the current structure of work itself can be discriminatory because it uses the image of the male worker as the metric for determining the characteristics of **the ideal worker**. Ideal workers, in today's economy, are people who can put in long hours, uninterrupted, throughout their careers. They are available when the company needs them, can work late when called on, can spend weekends in the office, and will not have any prolonged absences from the job. Their performance will not be hindered by sickness, aging parents, children, or pregnancy. These expectations are largely insensitive to the normal life courses of women and their roles off the job. In a word, ideal workers are men (see also Moen and Roehling 2005).

As another example, consider how inhospitable most workplaces are to breastfeeding—a normal part of women's lives, but not men's. Exhibit 6.10 reveals the numerous health benefits of breastfeeding, for both children and nursing mothers. The American Academy of Pediatrics recommends that newborn children be exclusively fed breast milk for the first 6 months of life, and then be provided with breast milk along with other food until at least 12 months of age. But fewer than one in four American women who work full time nurse their children to the 6-month marker, a much smaller percentage compared with new mothers who are out of the labor force (Galtry 2000, 2002).

The breastfeeding problem can be attributed to the fact that most working women cannot afford temporary leaves from their jobs and that newborn children are not welcomed in most workplaces. However, there exists an underutilized technological solution that can help women stay strongly attached to jobs. Breast pumps (which used to be bulky and somewhat noisy appliances) are now quiet and small enough to be carried like briefcases. These devices enable nursing mothers to express milk while away from their children and then store it for later use. Using these machines only requires privacy, access to a refrigerator, and the provision of brief breaks (20 to 30 minutes a few times a day). These are modest resources to provide, but they remain unavailable to most workers. In 2010, The Fair Labor Standards Act was amended to require employers to provide reasonable break time and a private, nonbathroom place for nursing mothers to express breast milk during the workday, for one year after the child's birth. Time will tell if this legislation has a marked effect, and as with other similar types of legislation, compliance is not always assured. Nonetheless,

Exhibit 6.10   Health Benefits to Breastfeeding

Benefits to Children	Benefits to Mothers
Less urinary tract infection	Less breast cancer
Less respiratory infection	Less ovarian cancer
Less diarrhea	Less osteoporosis
Less allergic disease	Earlier weight loss
Less middle ear infection	Enhanced self-esteem
Less bacterial meningitis	Enhanced infant bonding
Less botulism	Enhanced feelings of success
Less bacteremia	
Less gastrointestinal infection	
Less atopic eczema	

Source: American Academy of Pediatrics (2005).

this type of responsiveness to concerns central to the operation of the new economy provides hope, as it illustrates that expectations about workplace practices are shifting.

In sum, a focus on job designs and the standards used to evaluate employees reveals structural considerations that disadvantage women in the paid labor force. Effecting change in this area requires rethinking taken-for-granted assumptions concerning the best means to attach people to jobs and how those jobs are to be designed. This requires new talent management strategies to generate interest in jobs, screen applicants, design work spaces, construct work schedules, and evaluate employee performance. The first step in this process is to rethink who is the "ideal worker" and thereby bring workplace policy and practice into line with what is reasonable to expect from today's diverse workforce.

Strategies to Bridge the Care Gaps: International Comparisons

The increased participation of women in the paid labor force has contributed to resource gaps in the provision of care for children and aging parents. This problem will escalate in coming decades as the aging baby boom generation continues to transition into retirement and then into old age. As this happens, ever-larger proportions of the workforce will find themselves sandwiched between the need to work, care for their parents, and care for their children (Neal and Hammer 2006; Sarkesian and Gerstel 2004). Here we consider two core policy considerations—family care and family leave—and how those policies in the United States compare with those in other advanced societies.

America's approach to handling child care is a hodgepodge of stopgap approaches that create divergent experiences among different classes in the workforce. For workers at the top of the class structure, such as Erika, the stay-at-home mother discussed at the beginning of this chapter, high incomes afford not only the luxury to stay home, but also the resources to employ others to cook, clean, and at times watch children. Professional women with demanding jobs can purchase care privately, although usually at considerable expense, and rely on a variety of prepackaged care items, such as prepared foods. Some may be in a position to request reduced or flexible work schedules, especially if they are long-service employees and have well-paid spouses (Meiksins and Whalley 2002). These strategies work, but eat into couples' incomes and stall careers. They are also not the preferred arrangements, as many women feel deprived of the opportunity to engage fully in the provision of hands-on care of their children. For workers in jobs that pay low wages, the strains are far greater and solutions are even less satisfactory.[2] Many parents in working-class families deliberately work alternating shifts, so that someone is home at all times. This can work, but it has obvious negative effects on marriages and can create high levels of stress in people's lives (Presser 2003). It is not unusual for children in low-income families to be left to care for themselves (Heymann 2000). Note also that a full understanding of the problem with the American system of day care requires considering the ways that care needs are connected with class and racial relations. Today, America's day care system operates on the backs of poorly paid domestic and child care workers, who are almost exclusively women and disproportionately members of minority groups or recent immigrants. The affordable labor they provide makes it possible for professional women to devote time to their jobs, revealing that the ability of some women to integrate themselves into the new economy depends on the exploitation of other women (Mohanty 2003).

The uneven quality of day care in American society compares unfavorably with the situation in many countries in Western Europe, which have implemented national child care programs that provide high subsidies to fund publicly financed child care centers, which in turn promotes high use (Gornick and Meyers 2003; Pettit and Hook 2005). For example, because most child care workers in Sweden are well paid and have university degrees, virtually all working parents can go to their jobs assured that their children are receiving high-quality care (Morgan 2005). American culture embraces the notion that children are best raised in the warm arms of their mothers and considers day care providers to be comparatively "cold." This perception is not shared in Denmark, a country with universal child care supports. Danes believe that children served in quality state-sponsored centers receive "warm" care that helps them grow and mature, and it is not considered inferior to the care received in the home (Kremer 2006a).

Subsidizing care centers is one means of promoting family-responsive policies. But this does not address the fundamental problem that most care work is uncompensated and performed by family members (Crittenden 2001). Here another social experiment is enlightening—the provision of direct payments to individuals to purchase care assistance. In the United States, professional home health care assistance, such as in-home nursing, is available only to those who have financial resources. Those lacking resources commonly go without care or rely on family members to provide this care (which in turn can interfere with their work). In Denmark, those needing home assistance are given a "Personal Budget" that they can use to purchase the assistance of their own choosing. Interestingly, 64%

of the people employed by those controlling their own Personal Budgets are the spouses, parents, or adult children of the person purchasing care. In essence, this system enables family members to compensate each other financially for providing care work and legitimates their taking time out from the paid economy. However, this system is not without its own problems, because role conflicts can result when family members employ one another. It also has undermined the power of professional groups to define reasonable care standards (Kremer 2006b).

There have been some advances in the U.S. approach to work-family policy, but they are modest by international standards. One of these occurred in 1993 with the passage of the **Family and Medical Leave Act** (FMLA).[3] This legislation provides workers with the right to take 12 weeks of leave from their jobs to care for newborn and newly adopted children or to assist other family members in need. By recognizing workers' need to take "time-outs" to tend to family matters, the FMLA was a tremendously important step in helping address existing care deficits. But it has not helped all workers. Because the leave allowed is relatively short (12 weeks), the amount of time that care work can be performed is very limited. Employees who have worked for their employers for less than one year, those who work part time (less than 25 hours per week), and those who work for smaller employers (with fewer than 50 workers) are not covered by the FMLA. This disadvantages low-income workers, who are much more likely to be employed in these smaller enterprises, to have less stable job histories, and to work part time. Most importantly, because the leave is unpaid, it is useful only to those workers who can afford to take time away from their jobs. Ironically, the result is that those who most need to take time away from their jobs—those who cannot afford to pay for care services—are excluded from reaping the benefits of the FMLA (Gerstel and McGonanagle 1999). And when employers voluntarily offer the option for care leaves, they are more likely to be offered to men in upper tier professions than to women in lower tier occupations (Swanberg, Pitt-Catsouphes, and Drescher-Burke 2005).

The FMLA is family leave American-style; it assumes that most families have a breadwinner who can remain in the labor force to provide for family economic needs. For those eligible at the lower end of the economic spectrum, family leave can be considered a right, but not an option. Exhibit 6.11 on page 157 shows that the American approach is vastly inferior to the leave policies adopted in nearly all other developed nations. For example, Norway and the United Kingdom offer new parents an entire year off after the birth of a child, and these workers receive 80% or more of their base pay during that time. As we discuss next, provisions in Sweden are even more generous. In most of the countries that have paid parental leave, it is not the employer who pays (as Americans commonly assume would be the case). Rather, the government provides stipends (supported through higher taxes) that replace the wages new parents would have earned from their jobs. Though most Western European countries have developed more supportive family-responsive policies, important cultural variations also have influenced the forms these policies take. Some societies seek to emancipate women from the home and to dismantle traditional gender roles (e.g., Finland), while others seek to facilitate the fulfillment of traditional gender roles with some integration of women into the paid labor force (e.g., the Netherlands and Germany) (Haas and Rostgaard 2011; Pfau-Effinger 2004).

Sweden offers an interesting case study of the possibilities for policy change, as well as cautionary findings on the impact family-responsive legislation can have on men's and women's behavior inside and outside the workplace. Currently, families are entitled to a total of 480 days of family leave, most of which is compensated at 80% of the salary earned (with a cap on higher earning households). In the early 1970s, the Swedish government formed a commission to study family needs. Its findings pushed the nation to enact and actively promote family leave with legislation that was initially designed to encourage both mothers and fathers to be involved in the nurturing of children. It soon became clear that men, on the whole, were unwilling to make use of parental leave benefits. In response, the government initiated legislation and public relations campaigns to promote a "daddy month"—an incentive specifically directed to encourage men to take paid time out from their jobs to tend to their children. Interestingly, the impetus for this initiative to motivate men to take time out for work came first from women's groups, who saw rebalancing family divisions of labor as a critical condition for increasing gender equality. Later men became highly involved and helped win passage of the legislation (Bergman and Hobson 2002). Because the daddy month is defined such that one must either "use it or lose it," it has motivated men to take time out from work. Incentivizing men's participation in care work has been shown to be a critical factor in reshaping their commitments to family and the ways in which they attach themselves to work (Haas and Rostgaard 2011).

Sweden's experiments with family-responsive programs also helped reveal a paradox. While paid family leave facilitated women's participation in the labor force and helped them care for young children, it also *exacerbated* the level of gender occupational segregation and the earnings gap that separates men from women workers. The reason for this dynamic is that paid family leave is almost exclusively used by women (Morgan and Zippel 2003). When employers know that men are unlikely to request time away from the job, but that women are more likely to make that request, clear incentives are created to discriminate against women workers, especially

Exhibit 6.11   Family Leave Entitlements in Developed Countries

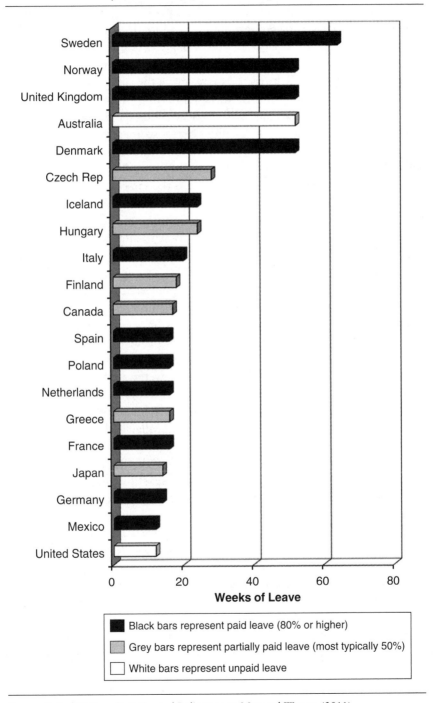

Weeks of Leave

- Black bars represent paid leave (80% or higher)
- Grey bars represent partially paid leave (most typically 50%)
- White bars represent unpaid leave

Source: United Nations Statistics and Indicators on Men and Women (2011).

those who will be expected to assume positions of responsibility. As a result, countries with the most family-friendly legislation (such as Sweden) have even higher levels of sex segregation than the United States does, and women are less likely to advance to the highest levels in the paid labor force (Mandel and Semyonov 2004, 2006). This paradoxical outcome—the more generous the family leave, the more severe the gender inequality—has intrigued researchers interested both in supporting gender equality and in resolving work-family tensions. What is clear is that the provision of family leave tends to expand gender inequalities for those in upper tier professions, but inequalities are leveled among those further down the occupational scale. Thus there needs to be caution in assuming that any work-family policy affects all classifications of workers in the same way (Mandel 2011; Misra, Budig, and Boeckman 2011). It is also clear that the type of work-family policy being considered, and the quantity of resources provided, can make a marked difference in shaping women's attachment to the labor force and men's attachment to care work (Hegewisch and Gornick 2011). Therefore, even if some unanticipated consequences are identified, these do not indicate a need to discard family leave policies, but rather to modify them—for example by introducing use incentives for men.

At a time when many European countries have been expanding their supports for working families, the United States enacted regressive policies directed at those at the bottom of the class structure. Ironically, as the need to care for disadvantaged children increased, welfare reform undermined the prospects for poor single mothers to stay home with their children. The Personal Responsibility and Work Opportunity Reconciliation Act of 1996 reworked the previous major welfare program, called **Aid to Families With Dependent Children** (AFDC) into the current program called **Temporary Assistance for Needy Families** (TANF). The AFDC system provided monthly welfare checks to poor families as a substitute for attachment to the labor force. Political leaders came to believe that this approach was promoting a culture of dependency, one in which poor people accepted "welfare as a way of life" (Murray 1995). In its stead, TANF was designed to limit support to a lifetime maximum of 5 years (states are free to set the limit lower, and some have done so), as well as to create an expectation that recipients work while receiving assistance and continue to work after the assistance ends. In this way, TANF was designed to get the poor into the labor force. But it did so with inadequate provisions to enable poor families (disproportionately single mothers) to manage the care of their children (Hays 2003). Studies of these families on the fault line indicate that they are typically placed in jobs that pay at or near the minimum wage, that these jobs seldom have any prospects for upward mobility, that care responsibilities interfere with women's abilities to keep these jobs, and that the lack of child care resources forces families to place children in situations untenable by middle-class standards (Crouter and Booth 2004). So, whereas many societies in Europe have enabled women to take sabbaticals from their jobs, America introduced initiatives to require the participation of the poor in jobs that pay little and offer few opportunities for advancement. Time will tell what impact this has on the next generation, but it is reasonable to predict that forcing poor women to take low-wage jobs will do little to alter existing gendered wage disparities and will maintain or even exacerbate racial disparities as well.

Conclusion

In this chapter, we surveyed the contours of gender inequality in the workplace, mapped the shape of these inequities, and considered the reasons for their occurrence. We showed that even though women comprise nearly half of the American workforce, only a small fraction of these women rise to the top and become corporate or government leaders. Instead, they are disproportionately funneled into lower-paying jobs and face greater prospects than men of having their careers dislodged midstream. We also showed that women perform substantial amounts of work that is economically unrecognized or undervalued. Although interpersonal discrimination remains an important consideration, deeper structural problems exist that differentially affect women's and men's careers. To be successful in today's workforce requires molding oneself into jobs designed for men who had stay-at-home wives.

The chasms that separate men's and women's careers are created by a variety of forces that have the consequence of fostering disadvantages for women that accumulate over their life courses. Women are coached early on to have different aspirations than men, and men and women are encouraged to accept—as a given—that "women's work" is of lesser economic value. If this work is undervalued, men will not seek it, with the result that large numbers of women will remain trapped in low-paying occupational ghettos. There have been advances in encouraging girls to aspire to enter "men's" careers. But there is a crying need to teach boys (and men) the rewards that can be attained in the performance of "women's work," in the home and in the paid economy, and to alter existing practices that systematically undervalue care work.

Although not a panacea or without problems and concerns, America's adjustment to the new economy can benefit from considering the far bolder European social experiments that advance opportunities to care for children and other community members. Ultimately, the future structure of the new economy will reflect the decisions made to address the question of gender inequality, the level of commitment to dismantling the gendered divisions of labor both inside and outside the home, and the concern for providing care when most parents are in the paid labor force.

Notes

1. The average height of American men is 69.2 inches, for women, 63.8 inches.

2. Parents in the United States can take a pretax deduction for child care expenses, but this policy does little to help low-income families who commonly pay little in taxes to begin with.

3. In California, legislation was passed in 2004 mandating that employees be eligible to receive as much as half pay for a period of 6 weeks to care for a sick or injured family member, or following birth, adoption, or foster care placement. Unlike the FMLA, there is no restriction on organizational size or tenure with the employer. Time will tell if the country follows California's lead, one that was met by considerable opposition from business groups (Nowicki 2003).

Chapter Seven

Race, Ethnicity, and Work

*Legacies of the Past, Problems
in the Present*

Jamal, the young African American fast-food employee introduced in Chapter 1, has the odds stacked against him. He grew up in a family that lacked resources and in a community that had few opportunities. But one wonders what factors caused his current situation, his tenuous hold on work in the new economy. Are his problems simply the result of his (and his parents') choices? Or are they the result of his lower-class social origins? No doubt both played a role. But, beyond that, the resources and opportunities available to Jamal have been influenced by race as well. As we discuss in this chapter, while some of the racial barriers that existed in the old economy have been partially countered by legislation and changing public attitudes, the structuring of the new economy also has introduced new challenges that disadvantage racial and ethnic minorities, which is in turn leaving a generation of young, lower-class minority workers at the margins. And even among members of the middle class, race continues to influence how high one can rise.

In this chapter, we offer an overview of race, ethnicity, and work, looking both to the past for the origins of contemporary problems and to the future challenge of creating a fully racially integrated workforce. We first consider race and ethnicity from a historical perspective to provide a context for considering the ways in which racial and ethnic exclusion manifested themselves in the old economy. Though patterns of exclusion have changed in some ways, they continue to influence the allocation of the resources needed to prepare for work, as well as access to jobs themselves. As a result, members of some racial and ethnic minorities are at a distinct disadvantage in today's workplace and labor market. These observations lead us to several critical policy issues concerning work, race, and ethnicity in the 21st century, including immigration control and affirmative action programs.

Histories of Race, Ethnicity, and Work

Race and ethnicity have been vitally important issues in the history of the United States. What is meant by these two terms, however, is not always entirely clear. Complicating the issue further, racial categories have changed dramatically over time. For example, Americans now refer to all persons of European descent as being "white," whereas early 20th century Americans would have categorized them into many different racial groups. Sociologists and others have increasingly come to define race as a social construct, in which physical differences between groups (such as skin color) are assumed by members of society to mean that those groups inherently differ in many other ways (e.g., ability, intelligence). The term *ethnicity* (or *ethnic groups*) is generally used to describe the cultural differences (e.g., language, diet, costume) between groups (for example, between Americans of Polish and Irish descent) (Omi and Winant 1986).[1]

African American Exceptionality

One question commonly posed is why African Americans, who have been in America nearly as long as white Americans (and much longer than many other ethnic groups) are economically so far behind. One explanation is that, unlike virtually all other immigrant groups, most people of African descent were brought forcibly to America, severed from family ties, and sold into bondage. In early America, African Americans were, in most instances, forbidden to hold property or to advance their positions. The dominant ethnic group, white Anglo-Saxon Protestants (WASPs), accrued wealth and later came to control the economic and political fortunes of the 19th century, but African Americans were left behind. Although American history books mark the resolution of the Civil War as the formal end to slavery, it did remarkably little to change the day-to-day work experiences of African Americans in the South, where most newly freed African Americans remained propertyless, sharecropped the fields for their old masters, and accrued debt rather than wealth.

Although a significant number of African Americans worked outside of agriculture even under slavery, it was difficult for them to find work in industry in the post–Civil War South. Some did find employment, but even those with significant skills were typically relegated to work as unskilled laborers. Whites and most trade unions (especially unions of skilled workers) devoted considerable energy to keeping African American workers out. And the racial hostilities that permeated American culture were strategically exploited by employers in some parts of the country, using African Americans as strikebreakers or threatening to replace low-wage white workers with African Americans unless workers acquiesced to management's terms. African American women were much more likely than were their white counterparts to work in the post–Civil War period, but they were largely kept out of industrial employment. Most found work in various menial service occupations, such as washing clothes (Gerstel and Sarkisian 2006; Harris 1982).

Around the time of World War I, large numbers of African Americans (along with rural whites from Appalachia) traveled from the South to the North to find work in the rapidly industrializing cities. Even though African American men succeeded in finding work in industry, most remained confined to unskilled jobs at the bottom of the employment ladder and were generally less likely to be employed than were new immigrants coming from Eastern and Western Europe. African American women had far less success entering industry than did African American men, so they continued to be concentrated in low-level service occupations. In the South, as well as in the North, a stable African American middle class emerged, but the black-white division was so strong that they formed separate castes within a supposedly free society, separated into distinct communities, and sharing few social ties or common experiences with the white middle class (Myrdal [1954] 1995).

In general, throughout the period of rapid industrialization, most African American workers, whether they were in the South or the North, were channeled into the least desirable forms of employment, were far more exposed to unemployment in bad economic times, and were prevented from acquiring the education or training that might gain them access to better-paying jobs. But there was some progress. Some unions, for example, particularly those that organized all categories of workers, not just skilled workers, organized across racial lines. And during World War II, because of labor shortages, some African American workers moved up, temporarily, into better jobs. However, once the war was over, many of these gains evaporated, leaving African Americans in much the same condition they had been in before (Harris 1982; Honey 1999).

In the latter half of the 20th century, many aspects of race in American society changed. Key pieces of civil rights legislation, such as the

Equal Pay Act of 1963 and the **Civil Rights Act of 1964**, established the legal standard that employers should not exclude anyone from employment on the basis of race or ethnicity. Civil rights legislation also put an end to legalized segregation in the South. Evidence also shows that overt hostility to African Americans has been on the wane and that the United States has been becoming a more racially tolerant society (Schuman et al. 1998). Does this mean that race no longer matters? Some, like the highly influential sociologist William Julius Wilson (1987) argued that race, as a master status, has declined in its significance. He observed that African Americans who had moved out of the central cities, who had entered college, and who had developed marketable skills stood reasonable chances of making it in the rapidly changing economy. If African Americans continued to be economically disadvantaged, he argued, it was not the result of race but of *class;* that is, of the declining opportunities for well-paid employment for those with limited education and skills. Others disagreed, finding evidence that race still mattered. Even though legalized discrimination had been abolished, and overt forms of prejudice weakened, the United States is still not a color-blind society, and various forms of "subtle discrimination" continue to affect all aspects of social life, including workplaces (Feagin 1991). As this chapter makes clear, good evidence supports *both* of these assessments.

The Immigrant Experience

African American workers faced particularly well-organized and harsh systems of oppression, but they were not alone in encountering obstacles to employment in the industrial United States. Americans frequently succumbed to "nativist" sentiments, and many immigrant groups found themselves excluded from certain kinds of employment ("Irish need not apply"), trapped in poorly paid ethnic economies (e.g., the Jews in the garment trades), exploited by unscrupulous employers, and suspected of radical or anti-American sentiments. In the factories of the late 19th century, jobs were distributed to ethnically homogenous work teams, and favoritism to one's own ethnic group was considered a normal way of distributing work (Jacoby 1991). Various stakeholders of the time produced a variety of documents demonstrating the dangers new immigrants posed to American society (Gusfield 1963; Thomas and Znaniecki 1958). As Exhibit 7.1 shows, these hostilities were compounded by concerns raised in the mid-19th century about Asian immigrants. This cartoon not so subtly suggests that immigrants, and their foreign ways, had the potential to swallow up American values and liberties. Racial and ethnicity diversity was then—and is now—commonly viewed as a threat rather than as a strength.

Exhibit 7.1   An Editorial Showing Fear of Immigrants (Circa 1860)

THE GREAT FEAR OF THE PERIOD
THAT UNCLE SAM MAY BE SWALLOWED BY FOREIGNERS.

Source: Copyright © by Corbis. Reprinted with permission.

During the 20th century, some hostilities faded as immigrants gained citizenship and assimilated. For the most part, immigrant families who had seemed foreign and threatening were amalgamated into the new economy, improving their fortunes within the expanding U.S. economy. But suspicion of immigrants remained strong, as evidenced by the passage of highly restrictive legislation that established strict quotas on immigration from virtually all parts of the world (Alba and Nee 2003). Certain groups received particularly unfriendly treatment, as the following examples illustrate.

- Male Chinese workers were imported to the United States, largely to help build the transcontinental railroad, but they were not allowed to bring along their families or become citizens. Once the railroad was completed, Americans reacted to the arrival of immigrants from China and Japan by pressing for legislation to keep them out. These efforts were successful; Chinese immigration was stopped with the 1882 Chinese Exclusion Act and Japanese immigration with the Gentleman's Agreement of 1907 and the Immigration Act of 1924.

- Japanese immigrants to the United States (Issei) succeeded as farmers in Western states and hoped that their children (Nisei) would gain acceptance into the American mainstream. However, despite doing well in school, few managed to find jobs outside the ethnic labor market before World War II. Isolated and disliked, Japanese Americans were easy targets for the internment policy of World War II (Takaki 2008).
- The Bracero Program, established to respond to labor shortages during World War II, established work opportunities for immigrants from Mexico. As "guest workers," Mexicans were allowed to enter the United States temporarily to work and then return to Mexico, but the program did not permit long-term residency and restricted their employment to low-level jobs American employers found difficult to fill (Alba and Nee 2003).

Changes in immigration law in the post–World War II period put an end to the exclusionary laws affecting Chinese and Japanese immigration and eliminated the immigration quotas, with their racial overtones that had shaped immigration policy since the 1920s. The Bracero Program, too, was ended in the mid-1960s, eliminating the special category of guest worker, to which many Mexican migrants had been confined. Recent immigration law reforms opened the door for increased levels of immigration. As a result, the percentage of American workers who are foreign-born is approaching the level reached at the peak of the pre–World War I immigration. In 1900, 20.3% of the labor force was foreign-born; after dropping for most of the 20th century, the percentage began to rise again in the 1980s and had reached 15.6% of the labor force by 2007 (Newburger and Gryn 2009). Moreover, levels of immigration to the United States in the late 20th and early 21st century rivaled those of earlier periods, with more than a million new *legal* permanent residents being admitted in many years (U.S. Department of Homeland Security 2010).

As levels of immigration rose, a sense developed that the most recent immigration waves were different. New immigration rules gave priority to those with educational credentials, which influenced the types of immigrants entering the country and their ability to enter the good jobs being cultivated in the new economy. As discussed later in this chapter, this dynamic can help explain why some recent immigrant groups have succeeded in ways that resident African American groups have not. However, because priority was given also to immigrants with close family in the United States, many legal immigrants continue to arrive with limited education and skills (Alba and Nee 2003). Many are concerned that these immigrants will become a burden on society or that they will bid down wages and displace more expensive American-born workers, an issue we discuss at greater length later in this chapter. Particular attention was paid to cultural issues, as Latino migrants were (falsely, as it turns out) believed to be resistant to learning English (Portes and Rumbaut 2001). Some also felt that, in an age of easy electronic communication and cheap travel, immigrants, particularly from nearby Mexico, would remain tied to their cultures of origin and would resist "Americanization" (Alba and Nee 2003).

In this context, hostility to immigrants has reemerged, as evidenced by attempts to build a fence along the U.S.-Mexican border, growing support for aggressive government efforts to detect and deport illegal immigrants and their children, and the emergence of vigilante groups such as the Minute Men, who patrol for illegal immigrants. The concerns expressed in Exhibit 7.1 are echoed by the thesis of politician and commentator Patrick Buchanan's (2006) book *State of Emergency.* The only difference is that, instead of focusing concern on the Chinese, this book directs hostility toward Mexican immigrants.

The Magnitude of Racial Inequality in the New Economy

It is hard to argue that race and ethnicity are no longer major factors in determining one's fate in the workplace after considering the extent of the differences between Latino, African American, and white American incomes. One way of measuring these disparities is to examine per capita earnings; that is, the total of all earnings within a racial group divided by total group membership. Exhibit 7.2 reveals that average incomes for men with incomes in the four major racial groups have remained largely stagnant during the past three decades, and have actually declined since 2000.[2] Exhibit 7.3 shows that averages for women with incomes have risen more rapidly and have not declined since 2000 (although women continue to trail men by significant margins). These exhibits also show the continued existence of income gaps between whites, Asian Americans, African Americans, and Latinos. For men, whites and Asian-Americans outearn African Americans and Latinos by a significant margin, and that margin has not narrowed appreciably over the past decade. The gaps among women wage-earners are considerably smaller and have not changed significantly in the past 10 years. Asian American women have the highest incomes, followed by white women. African American women's incomes are only slightly lower than white women's, while Latinas earn substantially less, on average, than women in any of the other three groups.[3]

The large gaps in men's incomes are only part of the story. Another important question is whether the jobs obtained by African Americans and Latinos are sufficient to bring families to a reasonable standard of living. Answering this requires shifting analysis from individuals to families, and considering all the sources of income a family possesses. Family structures vary across social groups, with significant consequences for their earnings. For instance, African American families are far more likely than are white families to be single-parent families relying on a single income. Latino and Asian American families, on the other hand, are

Exhibit 7.2   Mean Income of Men With Earnings: United States 1970–2008 (Adjusted to 2008 Dollars)

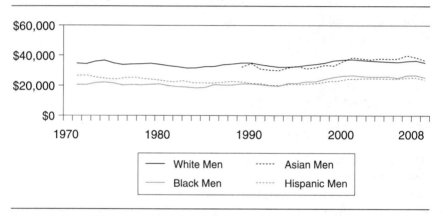

Source: Statistical Abstracts of the United States.

Exhibit 7.3   Mean Income of Women With Earnings: United States 1970–2008 (Adjusted to 2008 Dollars)

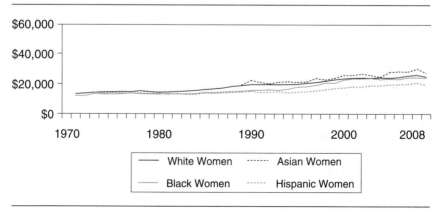

Source: Statistical Abstracts of the United States.

much more similar in structure to white families than they are to African American families. Having two adult wage earners in the household may help to compensate partially for the limited returns received from low-wage jobs. To deal with these complexities, we compare the proportions of *families* living below the poverty level in the four major ethnic groups. As Exhibit 7.4 shows, African American and Latino families are almost three times more likely than are white families to be living below the poverty line. This reveals that race and ethnicity are closely tied to the ability to find the types of jobs that enable families to achieve reasonable standards of living. For one in five African American or Latino families, these opportunities are very limited.

The existence of a significant income gap between whites and African Americans or Latinos has fueled speculation that a racially and ethnically distinctive "underclass" has emerged in the United States. This argument was first made about African Americans, whose higher rates of poverty and bad employment prospects are an old and (as the earlier data indicate) continuing story (Jencks and Peterson 1991). More recently, scholars have speculated that a similar phenomenon is developing among Latinos. These scholars argue that, unlike earlier European immigrants, Latinos are not rapidly assimilating into the American economy and moving their way up the economic ladder. Instead, a process of "segmented assimilation" is occurring, in which Latinos become part of the American economy, but in roles much like those of lower-class African Americans; that is, in poorly paid, unstable, secondary labor market jobs (Portes and Zhou 1993). Not everyone agrees with this view. Some contend that Latinos *are* gradually making their way up the ladder, although their progress is slowed by the

Exhibit 7.4   Percentages of Families Living Below the Poverty Line: United States 2008

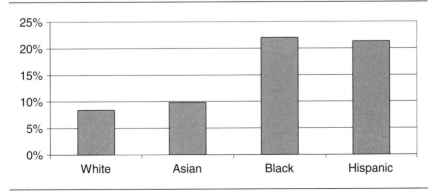

Source: Statistical Abstracts of the United States.

shrinkage of factory employment (the traditional route for upward mobility among immigrants) (Alba and Nee 2003). Still others argue that Latinos are experiencing what they call "working-class assimilation"; that is, they are moving up the ladder, but not as fast as they would if they were not members of an identifiable minority group (Waldinger, Lim, and Cort 2007). Though not fully agreeing on the pace of assimilation, all these studies conclude that being Latino has a significant negative effect on the ability to find good, secure, well-paying jobs.

Thus, we are left with a puzzle. How is it possible—in a society where the legal barriers to the employment of racial and ethnic minorities have been removed, where evidence indicates that some cultural attitudes have changed, where the belief is that equal opportunity exists, and where examples of successful minorities abound—that African Americans and Latinos lag so far behind whites and people of Asian descent? In the sections that follow, we argue that the reasons for racial and ethnic inequalities in the workplace are tied to three overarching sources of inequality, and that each of these sources of inequality involves persistent problems from the old economy and emerging problems linked to the new one. One source is the ability of parents to prepare their children for the world of work, a process that requires intergenerational transmission of resources. Another source involves aspects of social structure, such as the ways work opportunities, race, and geography intertwine. Finally, interpersonal dynamics play a role as well, as members of different ethnic groups face different prospects of being perceived and treated equitably in the workplace. As discussed later, these separate explanations—resources, opportunity, and discrimination—work together in a cumulative manner to fashion racial and ethnic disadvantages in the new economy. And while race and ethnicity create barriers for some members of society, they can create advantages for others by facilitating the formation of social connections that help members of some ethnic groups adapt to the new opportunity structures.

Intergenerational Transmission of Resources

For most workers in today's economy, a successful career requires strategic investments of time, energy, and financial resources. In this section, we consider four different types of resources that are passed through intergenerational relations. These resources constitute capital, or assets that can be used to enhance future career prospects (Bourdieu 1986). One type of capital is economic; that is, income and property that can be devoted to education or entrepreneurial endeavors. But prospects for success also can be elevated through the development of marketable skills (human capital), network ties (social capital), and the ability to interact comfortably with others who have resources (cultural capital).

Race, Ethnicity, and Economic Capital

In comparison with whites, African Americans and Latinos have less **economic capital**, fewer financial resources essential to pursuing a college education or the start-up funds needed for entrepreneurial endeavors. We have just seen that income is unequally distributed; so is wealth, the sum total of all possessions. Wealth is difficult to measure because it is not formally tracked in the way that income is assessed, but Exhibit 7.5 shows some of the extent of the differences. In 2004, fewer than one in two African Americans or Latino Americans had interest-earning bank accounts or owned their own homes. Fewer than one in 10 owned stocks, and only about one in four participated in 401K retirement accounts. Taken together, these statistics indicate that roughly one in two African Americans and Latino Americans is essentially propertyless, possessing virtually no economic resources to pass on to the next generation, either in the form of

Exhibit 7.5   Percentages of Individuals Owning Select Assets by Race: United States 2004

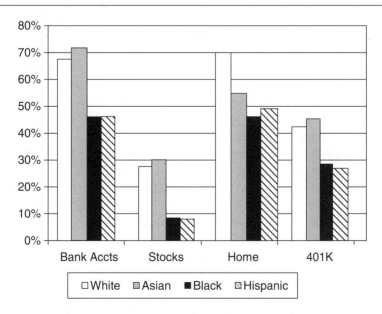

Source: U.S. Census Bureau.

inheritance or in the form of enhanced resources in the home. Note also that the absence of retirement accounts inverts the relationships between parents and children, as the children of African Americans and Latinos are far more likely to funnel resources *to* their parents in later life, which in turn saps the financial resources that they might send to their own children. The problem has only grown worse as a result of the ongoing housing crisis; African Americans and Latinos have lost their homes through foreclosure at almost twice the rate as white homeowners (Bocian, Li, and Ernst 2010).

Wealth deprivation is especially important to consider in the contemporary economic situation because it plays a significant role in decreasing the likelihood that disadvantaged children will successfully apply to competitive college programs. Opportunities to take advanced placement exams or to engage in extracurricular activities (necessary for entry into elite college programs) are in many cases absent from the rural and urban schools where African American and Latino students are concentrated. Movement to desirable school districts requires the financial wherewithal to afford down payments and hefty mortgages, often beyond the reach of disadvantaged parents (Kozol 2006). In other words, the possession of wealth (in the form of owning one's home) is directly linked to access to high-quality education.

How can the economic capital differences among racial and ethnic groups be reduced? One solution is to rework the distribution of resources so that members of the lower classes (who are disproportionately members of minority groups) have a more equitable share of the collective pool of income. This could involve, for instance, revising the minimum wage standard to enable those in low-wage jobs to earn enough to get by and to invest in their children's lives. Improved incomes would, in turn, help African Americans and Latinos begin to accumulate wealth (through savings, home ownership, and even investment), though efforts also would need to be made to overcome problems such as obstacles to home ownership (e.g., discriminatory lending practices and continued housing segregation) (Shapiro 2004). Finding ways to loosen the connection between property-ownership and access to high-quality education also would be important. This would require rethinking how community resources are directed; for instance, reversing the current system so that greater proportions of social resources are directed to schools in disadvantaged (minority) neighborhoods.

Race, Ethnicity, and Human Capital

At the beginning of the movie *The Graduate*, Mr. Robinson offers Benjamin Braddock a piece of advice about how to make his fortune— "Plastics . . . enough said." The message was straightforward; all Ben needed to do was develop the skills, apply himself to that industry, and his future would be made. Such advice fits well within **human capital theory**, a perspective that focuses on the value of the skills different employees apply to their jobs. **Human capital** comprises capabilities vested in individuals, such as technical skills, education, and experience. Those individuals who have the most human capital generally are the most valued employees, earn the highest wages, and have the strongest prospects for upward mobility. Conversely, those who lack skills, or who possess skills that do not fit current opportunity structures, experience economic hardship.

One approach to documenting differences in human capital is to study achievements in formal schooling. African Americans (84%) are earning high school degrees at rates that are only modestly lower than whites (87%) and Asians (88%), a dramatic change from the past, when African Americans were far less likely to complete high school than other groups. Latinos, on the other hand, continue to lag far behind (62%), a dynamic that largely reflects the recent arrival of poorly educated immigrants from Mexico and other parts of Latin America (Waldinger 2001). But even if high school graduation rates among the various ethnic groups were at complete parity, African Americans and Latinos are far more likely to attend schools that provide inferior learning opportunities compared with those offered in the suburban schools whites and Asians are more likely to attend.

More revealing are differences in the likelihood of completing postsecondary education; these trends are illustrated in Exhibit 7.6. While college completion rates have increased for all groups during the latter part of the 20th century, Asian Americans have the highest college completion rates, with whites second. African Americans and Latinos are significantly less likely to have a college degree, which puts members of these groups at a distinct disadvantage in the competition for the higher-skill and higher-paying positions in the new economy.

The racial differences in educational attainment help resolve one of the major questions concerning race and work: why are so many Asian Americans (many of whom are recent immigrants) succeeding when large numbers of African Americans and Latinos are not? One explanation can be found in Asian Americans' possession of the type of human capital valued in today's economy: the skills obtained through a successful college education. They are what some sociologists call "human capital" immigrants (such as computer programmers and doctors from India), who typically adapt to the United States quickly and get jobs fairly similar to those obtained by nonimmigrants (although there is evidence that Asian Americans with degrees from outside the United States do less well than those with degrees from American universities and that "silicon ceilings" exist in high-tech firms for Indians). Latinos are much more likely to be

Exhibit 7.6   Percentages Graduating College (Age 25 Years and Older)
by Race: United States 1970–2009

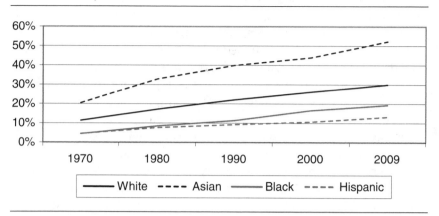

Source: Statistical Abstracts of the United States.

"labor migrants," who arrive with limited education and, consequently, make slower economic progress (Alba and Nee 2003; Varma 2006; Zeng and Xie 2004).

The strong effect of human capital differences on economic inequality points to the need to improve opportunities—especially for members of disadvantaged minority groups—for education and training. But this alone is not a solution to the problem of inequality because many jobs in today's economy are designed to be performed by low-skill workers. Even if one could wave a wand and make all members of society hold PhDs, there would still be hamburgers to flip, shelves to stock, and cash registers to operate. This indicates the need for initiatives to improve the returns on work for the disproportionate shares of minority workers who remain dependent on the bad jobs in the new economy. And, as Katherine Newman (2006) has argued, this will involve not just better wages, but also various kinds of support for working people with limited resources (e.g., child care, health care).

Race, Ethnicity, and Social Capital

Human capital is about "what you know"; in contrast, **social capital** is about "who you know." Social connections to wealth and opportunity are important to the process of getting a job and moving up in the labor market. People tend to get jobs through the people they know (Granovetter 1973, 1995). Imagine how much better off Benjamin Braddock would have been if Mr. Robinson had owned a plastics company!

African Americans, recent immigrants, and white Americans continue to interact in **homogamous networks,** or sets of social relations with others who are similar in race and economic position. African American and immigrant networks may help them get jobs, but those jobs are likely to be low-level jobs. Those at the lower end of the class structure are less likely to have much contact with people who can help them move up occupational ladders (Fernandez and Fernandez-Mateo 2006). Moreover, among the poorest groups in the new economy (where African Americans and Latinos are overrepresented), social networks often get used, out of necessity, primarily for day-to-day survival, and can even create demands on women (who are most frequently called on to help in times of need) that make stable employment unlikely (Dominguez and Watkins 2003; Stack 1997).

The existence of social capital plays an important role in explaining the successes and failures of immigrant groups to prosper in the new economy. For instance, when Nicaraguan professionals migrated to Florida in the 1980s, they experienced difficulty finding good jobs, despite their high levels of education. In contrast, relatively poor and uneducated groups such as the Vietnamese who settled in Louisiana during the same period tended to do better, despite possessing weak English skills. The key difference was the strength and extent of the Vietnamese family networks that helped new members learn of job opportunities and placed them in positions in locally owned businesses (Portes and Rumbaut 2001).

Other studies reveal similar effects for other ethnic groups that have had a longer history in the United States. For example, one study compared the employment experiences of young men growing up in three ethnic neighborhoods in Brooklyn, all of whom had high risk for entering criminal careers. When young Italian American men got into trouble, male relatives helped them get into the industrial work that they themselves performed. By contrast, Latino youths tended to work various kinds of neighborhood jobs with friends and relatives (especially those involving automobiles), which led them into more organized illegal work (in "chop shops") or into auto mechanic's jobs. African American youths' networks linked them mostly to various kinds of public sector employment where connections were of less help. These youths also had fewer network resources that could keep them out of trouble (Sullivan 1989). These findings indicated that the types of ties, and the strengths of social ties, mattered in the careers of these young men.

Access to networks and social capital is also an important factor in explaining the differences between African Americans and recent Latino immigrants. Both groups suffer a poverty of social capital, but in different ways. One problem for inner-city African American young people is that their teachers and family members have few social ties with employers and, as a consequence, have fewer doors opened for them (Kirschenmann and

Neckerman 1991; Moss and Tilly 2001; Royster 2003). They also have far fewer contacts with jobholders, and, as a result, have limited information about opportunities or mentorship on how to develop a successful career (Holzer 1991; W. Wilson 1997). Interestingly, Latino immigrants in inner-city locations appear to be finding jobs, despite the shortage of available opportunities, partly because of their social ties to other Latinos who hold jobs (Mouw 2002). These networks are so effective that some workplaces have developed internal norms of referring and hiring only Latino workers. Given language barriers, these tightly knit work groups can make it difficult for employers to hire members of other ethnic groups (Waldinger and Lichter 2003). In contrast, African Americans have been less able, or less inclined, to establish this kind of ethnic enterprise so their ability to help coethnics in this way has been far more limited (Butler and Herring 1991). Even though inner-city Latinos are more likely to be employed than are African Americans, their network ties tend to link them to relatively low-wage jobs (Fernandez and Fernandez-Mateo 2006; Waldinger 2001; Waldinger and Der-Martirosian 2001).

Differences in social capital also make enormous differences in chances for upward mobility within organizations. One's ability to move up the ladder depends not just on merit but also on mentoring and sponsorship received from senior employees who provide resources, opportunities, and protection when the going gets rough (e.g., Jackall 1989). Members of racial and ethnic minorities are at a disadvantage in getting this kind of help partly because of their scarce representation in positions of power, which creates a dearth of the kinds of social ties that are available to well-connected white protégés (Ibarra 1993). When members of minority groups *do* have access to strong, resource-rich networks, they are much more successful in gaining access to high-level jobs and senior positions (Ibarra 1993; Zweigenhaft and Domhoff 2003). As we discuss later, despite its many problems, this is one of the primary justifications for the continued application of affirmative action policies.

Race, Ethnicity, and Cultural Capital

The best jobs in the new economy require not only technical skills, but also the ability to work effectively with people. This requires knowing when and how to discuss issues, and how to comport oneself appropriately, especially when interacting with others who can open up opportunities. The focus on **cultural capital**—general cultural orientations and dispositions that can be used for gain—shows that it is not simply the ability to work, but also the ability to "fit in" that influences success (Bourdieu 1986). Acquiring cultural capital requires more than simply earning a college degree; it necessitates the ability to harmonize interaction, create friendships, know the right conversational skills, operate within taken-for-granted behavioral norms, and master other intangibles that are not part of formal job descriptions.

Culture operates like a toolkit, enabling individuals to navigate the worlds in which they live (Swidler 1986). But depending on the social world, the tools in that kit will vary. Members of racial and ethnic minority groups are more likely to come from families at the bottom of the economic ladder, so many have not been initiated into middle-class culture by either their families or the schools they attended. Consider, for instance, what it is like to be a young man growing up in an inner-city neighborhood where illegal drug use is prevalent. In this world, the code of the street is to gain and maintain "respect," and success in the neighborhood requires toughness and the ability to exert masculine dominance. But though these cultural skills facilitate survival in the neighborhood, they prove counter-productive to presenting oneself in ways that would be appropriate to work in downtown offices. When these young men venture into this world, they feel awkward and inadequate, aware that they do not fit in and that their language and attire is too "street." They also find it difficult to take orders or tolerate indignities, especially those they perceive as coming from women supervisors (Anderson 1999; Bourgois 1995). Nor is this simply a matter of their own attitudes; employers also look for the "right" cultural capital in prospective employees. Unconventional (from the middle-class perspective) forms of dress or appearance (e.g., braids, dreadlocks, beaded hair), non-standard English, and other characteristics of inner-city African American life are often looked down on by employers, who see these as markers of unreliability and an inability to fit in (Moss and Tilly 2001).

Cultural capital is not learned in the same way one might learn computer programming. It comes from years of socialization that form a **habitus**—the habitual ways of interacting with and understanding one's world (Bourdieu 1984). This requires more than memorizing whether forks go on the left or the right. It requires the development of a natural ease in the situations that matter and the ability to belong with those who count. And it involves developing perspectives about what to expect for oneself and what to expect from others. One study of successful African Americans found that they stressed the need to learn to fit into the social world of more affluent whites. These African Americans argued that their experiences in private schools had taught them to be confident and comfortable in a social environment very different from the one into which they were born (Zweigenhaft and Domhoff 2003).

Lower-class minority youths tend to be very different, in terms of cultural capital, from successful African Americans and successful whites.

Classic studies of lower-class young men indicate that they often lack a clear sense of what types of jobs are available, may have unrealistic aspirations, and often develop a perspective of self-blame when things do not work out as they had hoped (Liebow 1967; MacLeod 1995). A more recent study of young African American men living in Chicago found that they held fairly conventional (if often unrealistic) goals in a community where it was unlikely that they would find good work. They wanted "good jobs" that brought respect; they talked about city jobs, starting small businesses, and, notably, getting blue-collar jobs in factories, even though few such jobs remained in the communities in which they lived. However, when asked what they *expected* to be doing in a few years' time, they had no clear sense of the future or of how their aspirations would pan out. They possessed only a sketchy knowledge of efforts that might benefit them in obtaining work in the contemporary economy. In sum, they thought they could succeed but lacked the cultural capital to know how (Young 1999, 2003).

The importance of cultural capital in an economy where most work involves social interaction should not be underestimated. But how does one teach culture? Part of the problem is the extent of economic segregation, which has increased since 1970. As long as the social networks of those from poor backgrounds are restricted to others in similar situations, and as long as community environments promote different cultural codes and values, this will remain an enduring problem.

Geographic Distribution of Race and Work Opportunity

Understanding the interaction of race and work requires considering residential patterns, the geographic distribution of work, and the prospect that individuals can find good jobs that match the skills they possess. Although proximity does not guarantee that one will get a job, the chances of finding a good job will be better if jobs are located near one's home and community, and if one resides next to other people who have jobs. The **spatial mismatch hypothesis**—that is, the idea that people often have trouble getting jobs if they are located where employment is scarce—has been shown by many studies to help explain why minority workers (who are often spatially concentrated) have limited access to good employment (Fernandez and Su 2004). One of the most striking examples of how geography has affected the economic experiences of an ethnic group can be found in the case of Native Americans. The reservation system, established in the 19th century, virtually ensured that indigenous communities, established in remote rural areas, would be at the margins of the economy. The lack of work is one of the primary causes for this ethnic group's escalated risk for suicide, alcoholism, and out-of-wedlock births (Erikson 1994; Snipp and Summers 1992).

William Julius Wilson (1987, 1997) demonstrated comparable effects on the experiences of African Americans in the Northeast and Midwest. In the middle of the 20th century, when the industrial economies of cities like Detroit, Cleveland, and Chicago were booming, African American men found work in factories, and their families created stable communities. However, from the 1970s to the 1990s, those factory jobs disappeared, followed by the middle class, which left many neighborhoods in central cities populated by relatively low-income African American populations (Bartlett and Steele 1992; Bensman and Lynch 1987; Bluestone and Harrison 1982). The same types of economic and demographic shifts also occurred across much of rural America (Duncan 1992). In many of these places, a "hyperghettoization" of concentrated poverty resulted in increased crime, drug use, and out-of-wedlock births, which in turn undermined the prospects of these communities' nurturing new workers or attracting new employers (Sampson, Morenoff, and Gannon-Rowley 2002; Sampson, Raudenbush, and Earls 1997). The effects of industrial decline on African Americans do seem to have varied across the country. Waldinger (1996) for example, found that manufacturing was not so important to African American employment in New York City, so its decline there did not have the same racial consequences as it did in cities like Chicago. Others have noted that Latinos in Chicago were also highly concentrated in manufacturing, but did not suffer the same consequences as African Americans when industry there declined (Parks 2011). Still, the segregation of the African American population in job-poor communities is an important obstacle to African American employment.

Part of the challenge for the new economy is shifting work opportunities to the communities that are currently work-poor and facilitating the movement of families to locations where jobs are more plentiful (Newman 2006). The Gautreaux Program in Chicago illustrates what can happen when people are moved to employment-rich communities. It was designed to move some residents of public housing in the city of Chicago into private sector housing, and thereby reduce the impact of concentrated poverty on the lives of the disadvantaged (W. Wilson, 1997). Half of the participants were randomly selected to move into suburban housing, and the other half were moved into housing within the city of Chicago. Those who had moved to the suburbs were significantly more likely to become employed. There were various reasons for this, including the transplants' sense that the neighborhood was safer (so leaving their kids alone was less of a problem) and that *not* working was socially unacceptable. But perhaps the most powerful reason for the change in work outcomes was that the suburbs offered more opportunity to work and that the African American transplants had little trouble finding jobs for which they were qualified (something that could not be said for those who remained in the city). Other studies have shown that moving to suburban locations does not, by itself, always improve the employment prospects of African Americans (Fernandez and Su 2004). But the Gautreaux experience provides strong evidence that it *can*.

If many African Americans are disadvantaged by living in declining inner-city neighborhoods where employment opportunities are scarce, what about immigrants who also are likely to settle in urban areas? Here, the nature of ethnic communities produces complex, perhaps contradictory effects. In some cases, ethnic communities provide opportunities for migrants to find work in businesses owned by (or at least dominated by) others from the same ethnic group. This can provide immigrants with better jobs than they would have found outside the ethnic economy (Wilson and Portes 1980). For example, Cubans who migrated to the Miami area established a significant number of businesses, most of which were small, but that employed many of the Cuban Americans who settled there. Those Cuban Americans who worked in the Cuban "enclave" were not all trapped in low-paying jobs. The ethnic enclave provided significant opportunities for many to learn new skills, including business skills, and open their own businesses, serving as an avenue to economic success for at least some immigrants.

In other cases, working for or with coethnics has been a real disadvantage in the longer term, trapping the immigrant in a subculture and limiting opportunities to move into the economic mainstream (Alba and Nee 2003; Waldinger and Der-Martirosian 2001). For example, Asian immigrants in the Los Angeles area benefited from their ethnic enclave, at least at first, because it provided them with employment and other benefits when their language and cultural skills were still poor. However, the kinds of jobs they found in the ethnic economy typically provided low wages. If they remained within the ethnic enclave, they became trapped in low-wage undesirable employment, unless they were among the few who successfully established businesses. Fortunately, Asian immigrants have been largely successful, over time, in moving out into the larger economy as they become more acculturated. Their children, in turn, are even more likely to enjoy upward mobility and improved wages as a result (Nee, Sanders, and Sernau 1994).

People work in jobs, but live in communities. Ethnicity plays a role in determining where people choose to live, and ethnic enclaves create unique dynamics in shaping what types of goods and services are desired, which in turn influences the types of work opportunities that will be created. However, the movement of jobs into and out of areas where ethnic minorities are concentrated can have a profound impact on current and future generations of minority workers. As William J. Wilson (1997) found, when work disappears, it sets the stage for a wide variety of social problems to emerge. The solution, from a work opportunity perspective, is to redistribute work throughout the various communities, and to consider not just the racial characteristics of workers, but also the patterns of racial and ethnic concentration in neighborhoods and cities.

Racial Prejudice and Discrimination

The creation of a truly inclusive society requires that there be opportunities to work, that individuals have the resources to develop the skills to gain employment, and that once they are employed they are treated in a fair and equitable manner. Here, we consider the extent to which members of racial and ethnic minorities continue to experience employment discrimination.

Officially, employment discrimination is illegal in the United States. Laws passed in the civil rights era stipulate that employers should not discriminate against potential or current employees on the basis of their race (or gender), and that has resulted in a genuine decline in overt forms of racial exclusion (e.g., the deliberate exclusion of African Americans from apprenticeship programs or professional training programs). But the research evidence indicates that race has not been entirely eliminated from employers' decisions about whom to hire and promote. In fact, employers are surprisingly willing to admit to generalizing about members of racial groups and to express a preference for (or reluctance to hire) members of different racial groups as employees. Though not all feel this way, many employers still have strongly negative—and often hostile—opinions of African American employees, particularly African American men, many of whom employers regard as unreliable and difficult employees. The harshest appraisals are directed at African Americans from the inner city—that is, those who live in "bad" neighborhoods, attended urban public schools, and speak "Black English" and dress distinctively. Employers associate inner-city residents with a variety of negative behaviors (e.g., drug use, crime, poor education) and express reluctance to hire people from those neighborhoods. Employers respond more negatively to job applicants with inner-city addresses or with African American–sounding names, like Jamal or Lakisha (Bertrand and Mullainathan 2004; Fernandez and Su 2004; Kirschenmann and Neckerman 1991).

The strength and persistence of employers' reluctance to hire African Americans is well-documented in recent research by Devah Pager. Her research involved an experimental design in which matched pairs of job applicants (chosen to be similar in most respects other than their race or ethnicity) applied for entry-level jobs, then were tracked to determine whether race or ethnicity affected the outcome. As in previous research of this type, Pager found that whites received more positive outcomes than African Americans, with Latinos somewhere in the middle. Not only were African Americans less likely to get called back after an initial application, but they also were "channeled" into lower-status positions, while whites were more likely to be "channeled" toward better positions than the ones for which they had applied (Pager, Western, and Bonikowski 2009). Remarkably, Pager found that employers were more likely to react positively to white applicants with a felony conviction on their record than they were to comparable African American or Latino applicants who did not have a criminal record (Pager 2003).

Not only does discrimination play a role in decisions about whether to hire minority workers, but also in their experiences once they have been hired. Analysis of verified racial discrimination cases in Ohio found that the most common complaint in such cases concerned dismissal. The records contained numerous cases of African American workers who complained, successfully, that they had been dismissed unfairly. These workers reported various forms of discriminatory treatment at work, including receiving less help on the job, being assigned different tasks than coworkers, and being subjected to harsher penalties than white workers for various workplace infractions. Unfairly dismissed African American workers also often complained that, prior to their dismissal, they had been subject to various forms of harassment by both their managers and coworkers (Mong and Roscigno 2010).

Negative views of African American employees seem to focus particularly on African American men. However, a somewhat different (but also negative) perception of African American women has consequences for their chances of finding good jobs. Employers do not think of African American women as lazy or "difficult" in the way that employers tend to stereotype African American men (Browne and Kennelly 1999; Kennelly 1999). Actually, employers are more likely to perceive African American women positively, in some respects, emphasizing their role as breadwinners for their families and their role as parents. Unfortunately, this is combined with a tendency to stereotype African American women as single mothers and to project onto them problems that employers imagine go along with this status. For example, employers express concerns about tardiness and frequent absences from work in connection with African American women, believing that single parenthood will create complex work-family conflicts that cannot be resolved. Though many African American women are *not* single mothers, and though married women *and* men experience problems associated with work-family conflicts, employers continue to perceive African American women as single mothers and to have concerns about their value as employees as a result.

For many entry-level jobs in particular, employers place a great deal of emphasis on applicants' "soft skills," including communication and people skills as well as qualities such as "attitude" and "personality" (Moss and Tilly 2001). Employers generally have a negative view of African Americans' soft skills, believing that they are less reliable, more likely to "have an attitude" and experience difficulties interacting with members of other racial groups. As a result, employers often are reluctant to hire African Americans for jobs that demand soft skills. In some cases, employers try to exclude African Americans altogether; in others, employers try to "match" the characteristics of the employee to the racial characteristics of the workplace or clientele (e.g., by choosing to hire African Americans for settings in which they work with or serve other African Americans). The result is different, and diminished, employment opportunities for African Americans.

Recent increases in Latino immigration (both legal and illegal) have led to questions about whether employers prefer Latino employees to African Americans. There has been some suspicion that employers are choosing to hire Latinos for jobs once held by African Americans, such that Latinos are benefiting economically at the expense of African Americans. But the research evidence suggests that the reality is more complex. Employers perceive Latino workers as hardworking and deferential (Waldinger and Lichter 2003). Given the prevalence of negative views of African Americans as workers, it would not be surprising to find that employers prefer Latinos for jobs where skill demands are low and the main requirements are hard work and subservience. That these preferences exist, however, does not mean that Latino workers are displacing African American workers, as some suspect. To begin with, employers' preferences vary according to the type of position they are trying to fill (Waldinger and Lichter 2003). When skills or human capital are not an issue, the preference for Latinos takes precedence. However, for other jobs, Latinos may be less desired. When jobs require English-language skills, employers may actually *prefer* African Americans. Jobs in health care, for example, require the ability to interact with highly educated English speakers (doctors and sometimes patients), to read instructions on technically sophisticated equipment and materials, and to understand safety directions. Moreover, even where Latinos are engaged in jobs that once were performed by African Americans (e.g., in light manufacturing), one cannot assume displacement has occurred. What often is happening is that African Americans have voluntarily abandoned these jobs in preference for better positions that have become available (Lim 2001). The willingness of some immigrants to accept low wages reduces the pressure on employers to pay more (which would make these jobs more attractive to other ethnic groups). But the impression that low-wage immigrants "push out" African Americans (who then become unemployed) may be an oversimplification.

Eliminating stereotypical views of racial and ethnic groups is not something that the passage of a law or the implementation of a sensitivity program alone can achieve. However, developing and enforcing effective policies for combating the discrimination that results from prejudice *does* increase opportunity (Stainback, Robinson, and Tomaskovic-Devey 2005). And the belief that people cannot do something tends to weaken when they demonstrate that they can.

Racialized Jobs

Another facet of the effect that prejudicial attitudes have on patterns of employment is the existence of **racialized jobs,** or those jobs that are typecast to be held by members of specific racial or ethnic groups. One familiar example of this phenomenon can be seen in personal service jobs, such as those held by maids who perform work within other people's households. There is a long tradition in the United States of expecting African American and Latina women to perform domestic labor for others. Racial attitudes have changed, but race still affects maids' work experiences. African American and Latina maids are much more likely than are whites to be asked by employers to perform personal services, to engage in stigmatizing tasks (changing bedclothes, washing underwear, scrubbing on one's hands and knees), and to experience a variety of intentional or unintentional slights at work (Wrigley 1999).

Racialized jobs and work roles also can be seen in higher-level occupations. The case of the growing African American middle class is illustrative. In the past, the African American middle class tended to be confined to restricted niches. African American professionals, business owners, and the like existed, but they were largely concentrated in African American neighborhoods, providing services to African American customers and clients. In the past few decades, middle-class African Americans have been finding their way into predominately white organizations, especially in the public sector. In fact, some have argued that public sector employment has been more important to the economic fortunes of African Americans than manufacturing (which was important largely in Midwestern cities such as Chicago and Detroit). In this view, the expansion of public sector employment, particularly in the 1960s and 1970s, provided paths to upward mobility for many African Americans. However, the concentration of middle-class African Americans in public sector jobs has made them vulnerable. Recent cuts in the public sector have undermined previous gains, and African American men in particular have lost access to some of the well-paid government jobs previously available to them (Parks 2011; Waldinger 1996).

The contemporary African American middle class also suffers from other kinds of "ghettoization" at work, and studies have shown that African American managers and professionals are being funneled into smaller, peripheral firms (Vallas 2003b), have less authority (Elliott and Smith 2001), earn less even when they are in positions of authority (Smith 1997), and are more likely to supervise African Americans than whites (Elliott and Smith 2001). Evidently, being an African American professional or manager is not exactly the same experience as being a white one. Findings such as these indicate that many African Americans who enter the corporate world encounter a contradictory reality (Collins 1997). On the one hand, a desire to integrate workplaces, combined with affirmative action programs, makes minorities highly desired, and companies often actively seek these types of employees. However, even though the obstacles to entering the corporate world may have been reduced, what happens afterward is different. Many African Americans find that, once they have been hired, their progress up the corporate ladder is slow. It seems to take longer to get promoted, and they suspect that they have to meet a higher standard than do their white counterparts with similar kinds of education and experience (G. Wilson 1997). They also find themselves "steered" toward dead-end jobs in areas such as community relations and diversity offices, which offer attractive starting pay and a chance to "give back" to the community, but provide few avenues for career growth (Collins 1997).

Race, Ethnicity, and Work: Social Policy

The persistence of racial and ethnic inequality at work, and the tensions and conflicts surrounding its resolution, raise a host of policy issues. Earlier, we examined the dominant theories regarding why some racial and ethnic groups lag behind and why others are advancing in the new economy. For each theory, we highlighted the implications for policy reform. Here we consider two pressing policy debates: the continued use of affirmative action programs and changes in the United States' approach to both legal and illegal immigration.

Affirmative Action

Affirmative action is the most visible effort in the past 40 years to confront racial (and gender) inequality in the workplace. In 1968, the federal government began requiring firms with significant government contracts

($50,000 or more) and at least 50 employees to produce annual affirmative action reports and to demonstrate a commitment to hiring minority workers (including women). This became the basis of the programs we now know as affirmative action, which have been subsequently adopted by many companies, as well as by colleges and universities in their selection and recruitment of students (Stainback, Robinson, and Tomaskovic-Devey 2005).

How does affirmative action work? In most cases, the policy simply requires that employers make a "good faith effort" to recruit and retain employees from underrepresented groups—specific minorities and women. This means that employers must advertise vacancies in ways visible to minority candidates, explain why they did not hire minority candidates if qualified candidates applied for the positions, and produce regular affirmative action reports documenting their efforts.

Opponents of affirmative action label it as a quota system and conclude that it leads to the hiring of unqualified individuals. They also express concerns that affirmative action leads to reverse discrimination, whereby minority candidates are favored over whites in hiring and promotion decisions. Individuals not favored by the program commonly conclude that difficulties experienced in securing work result from opportunities being unfairly directed to minorities and women. Even some minority group members express concern that they are stigmatized by assumptions that they have been hired only because of their skin color (Steele 1990). National opinion polls, such as the General Social Survey, reveal that most whites do not favor affirmative action and that the African American community is divided on the issue.

On the other hand, proponents argue that affirmative action does not operate on the basis of quotas because specific positions are not set aside for minority candidates. Proponents also argue that affirmative action does not undermine the appointment of qualified individuals. To be hired under affirmative action programs, all candidates must meet base level qualifications. Once those criteria are met, race and gender are considered for two reasons. First, it is in the interests of society to distribute opportunities more equally, and this system (despite its flaws) is needed to combat discrimination and other factors that limit opportunity. Second, creating a diverse workforce or student body enhances organizational functioning in a multicultural society. National survey data indicate that although Americans generally oppose policies called "affirmative action," they are actually sympathetic to efforts to seek out and retain minority employees (Bowen and Bok 1998; Reskin 1998).

To date, the courts have upheld the constitutionality of affirmative action. But they have put constraints on how it is practiced by ruling that seniority rules take precedence over affirmative action (*Firefighters v. Stotts*, 467 U.S. 561 [1984]), by shifting the burden of proof in claims of discrimination to the person making the complaint (which makes it much more difficult to prove) (*Ward's Cove Packing Co. v. Antonio*, 490 U.S. 642 [1989]), and by making it easier for employees to challenge race-based hiring practices (*Martin v. Wilks*, 490 U.S. 755 [1989]). The recent Supreme Court decision regarding the University of Michigan's admissions policies further restricted affirmative action (*Gratz v. Bollinger*, 539 U.S. 244 [2003]). While the court ruled that race *could* be considered as part of a holistic approach to admissions decisions, it prohibited the university from continuing to use a quantitative rating system in which applicants were awarded points simply for minority status.

One of the most important questions today concerns the effectiveness of affirmative action programs—are they opening doors that would otherwise be closed? Strong evidence indicates that affirmative action policies have helped African American and women workers, especially those from the middle class, by pressuring major employers to alter their hiring practices (Reskin 1998). The most visible effects of affirmative action have been in the public sector, where minority employment has increased significantly, partly because of the greater effectiveness of enforcement. However, the effects have been less noticeable in recent years, as the political desire to enforce them has eroded (Stainback, Robinson, and Tomaskovic-Devey 2005).

There are limits to the success of affirmative action. One concern is that the programs primarily help African Americans, Latinos, and women who possess college degrees because they force employers to consider them for higher skilled jobs. In contrast, affirmative action has been of limited help to people at the bottom of the class structure, those who lack the qualifications for desirable employment. Additionally, although affirmative action opens some doors into workplaces, it does not combat the glass ceilings, "racialized jobs," and dead ends that minority and female employees often encounter after they enter large organizations and attempt to move up the ladder (Collins 1997).

Immigration Policy

A second policy issue that has dominated the recent headlines concerns what to do about immigration. Rising numbers of immigrants have led to concern about their economic impact and to calls for a return to a more restrictive approach. Much of the concern focuses on illegal immigration, but many Americans seem worried, especially in a period of stagnating wages and rising insecurity, that immigration *in general* has gotten out of hand.

A primary source of hostility to immigration is the belief that immigrants compete with American citizens for jobs. Economists and others have argued long and hard about the economic effects of immigration. Some conclude that the effects are largely negative, at least for American workers. They argue that an increase in immigration tends to result in lower wages because immigrants are willing to work "for less" and are vulnerable to unscrupulous employers. In addition, the growth in the supply of workers reduces the pressure on employers to raise wages to attract employees. Others argue that the situation is more complex. Areas with high rates of immigration (including *illegal* immigration) are not atypical in their wage structures (Broder 2006; Lowenstein 2006). Also, the argument that poorly educated immigrants compete with American workers is weakened by the fact that the vast majority of American workers have an education at the high school level or beyond, and so face only a limited threat of competition from such immigrants.

Analyses of local labor markets indicate that the effects of immigration on wages in particular areas tend to be slight (Sassen 1995). Some have argued that the real effect is at the national level; they contend that if immigration had not occurred, the wages earned by low-skill workers in particular would have been significantly higher (Borjas et al. 1997). However, these studies are inconclusive, and some economists conclude that the safest assessment is that immigration has had only a modest negative impact on the wage growth of low-skill individuals in the United States (and a somewhat positive effect on others' wages) (Card 2001; Saiz 2003).

Other critics suggest that the problem with immigration is that it results in higher unemployment. Immigration increases the labor supply, creating more competition for jobs, and may encourage employers to replace Americans with less expensive immigrant labor. The evidence in support of this argument is *less* persuasive than on the issue of wages. First, it is difficult to disentangle the economic effects of immigration from the effects of other economic changes; for example, if unemployment rises in an area that has experienced both immigration *and* plant closings, which of the two is responsible? The availability of low-cost immigrant workers also may cause employers to expand employment (since they can afford additional workers) so hiring immigrants may not displace any one at all (Porter 2006). In addition, immigration is often associated with economic growth, so even if workers are displaced by new migrants, an overall expansion of employment may create new job opportunities for those who have been displaced. A classic study of a sudden increase in immigration in a local labor market (the Mariel immigration of Cubans to South Florida in the late 1980s) found that the labor market adapted to the influx of immigrants, absorbed them successfully, and that unemployment was unaffected

(Card 1990). Workers, particularly immigrants themselves, also move when employment prospects are limited in a particular area, so an influx of immigrants does not typically lead to a growth in unemployment (Saiz 2003).

The issue of high-skill immigrants has excited controversy as well. Under current immigration law, employers are allowed to sponsor a number of immigrants from overseas on the grounds that they possess scarce skills that the employer requires and has difficulty obtaining in the U.S. labor market. The H-1B program allows employers to bring in skilled employees for temporary employment (though the employer also can choose to sponsor their application for permanent residency in the United States); the L-1 program allows international companies to "import" skilled employees who work for one of their overseas offices. As a result of such programs, the number of foreign-born workers with expertise in science, technology, engineering, and mathematics in the United States has increased significantly in recent decades (Lowell 2010). Defenders of these programs argue that they help the American economy by bringing in talented people from overseas and by enabling employers to fill vacancies that would have otherwise gone unfilled. However, critics argue that the programs are being abused extensively. One such criticism holds that the employees being brought in are not, in fact, highly skilled and that they are not paid the appropriate wages for the positions they occupy (since the employer holds the visa, the employee is vulnerable to such treatment). In this view, these programs undercut American workers' wages and create unnecessary unemployment and underemployment among native-born technically skilled employees. In addition, some of the heavier users of these visas are companies that critics claim are major outsourcers of high-skill work—foreign workers are being brought to the United States to receive training and experience, then return overseas where they perform jobs that used to be or could be done in the United States (Hira 2010a; Lowell 2010).

If the effects of legal immigration are complex, what about *illegal* immigration? The United States, has been experiencing a high level of illegal immigration, primarily (but not exclusively) from Mexico and other Latin American countries. While it is difficult to get an accurate count of the number of illegal immigrants in the United States, government estimates are that their numbers were as high as 11.8 million in 2007, although they have dropped off slightly to 10.8 million in 2010. Over half of this population is concentrated in five states: California, Florida, Texas, Illinois, and Arizona (Hoefer, Rytina, and Baker 2011). The growth of immigration from Latin America has been fueled by a variety of causes. The Bracero Program introduced many Mexican families and communities to the American labor market as temporary guest workers, but when the program ended in 1964, the immigration continued in different forms. Changes in immigration law in the 1960s made it easier for those in the United States to sponsor relatives as immigrants. And changes in the economy of Mexico (particularly in agriculture, where traditional small farmers confronted more-efficient, better-capitalized farms producing for global markets) pushed more Mexicans to leave their communities or to send family members away in search of income (Alba and Nee 2003). Illegal immigration has grown, in part, because changes in U.S. immigration policy have gradually made it more difficult for relatively unskilled Latin American migrants to enter the United States legally. The attraction of better-paying employment in the United States has been sufficiently powerful to overcome the dangers of entering the United States illegally and risking detection and deportation. The passage of the 1986 Immigration Reform and Control Act attempted to restrict the flow of illegal immigrants through a combination of increased border patrols and stricter sanctions on employers, who were made subject to penalties for hiring undocumented workers. During the administration of George W. Bush, it was rare for employers to be fined for employing undocumented workers, and only of late has the Obama administration changed tactics by shifting the focus from punishing the undocumented workers to the punishment of their employers (Preston 2011). This change in response may prove more effective because evidence is scarce that increased apprehension of illegal immigrants has discouraged these individuals from making an initial attempt, or from trying again (Cornelius 2005; Massey 2006). Moreover, focusing on interdiction at the border ignores the fact that many undocumented workers enter the country legally, on short-term visas, and simply stay on once their visas have expired. Building walls between Mexico and the United States does little to affect this aspect of the problem.

In 2010, Arizona passed legislation that attempted to go beyond interdiction at the border and a number of other states have begun discussing similar initiatives. Arizona's law focused on improving law enforcement officials' ability to detect undocumented immigrants by giving police officers broad powers to detain people suspected of being in the country illegally and by making it a criminal offense not to carry immigration documents. The law has been challenged in the courts and has excited considerable controversy, as it likely will expand the use of racial profiling. Nevertheless, laws such as this one reflect growing public concern about illegal immigration (Archibold 2010).

The most urgent question regarding illegal immigration concerns its effects on the United States, its economy, and its workers. As Exhibit 7.7 shows, illegal immigrants compose sizable proportions of the labor force in jobs that require low-skill work, especially in farming, cleaning, construction, and food preparation, where between 5% and 25% of workers are illegal immigrants.

Exhibit 7.7   Percentages of Occupations Filled by Illegal Immigrants: United States 2004

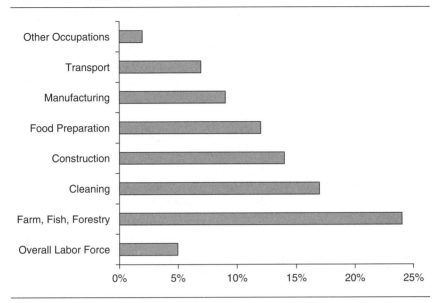

Source: Broder (2006).

Illegal immigration is believed to allow large numbers of undocumented aliens to enter the United States, drain public resources, take away jobs from legal residents of the United States, and exert a downward pressure on wages and working conditions (Buchanan 2006). Others object to these assumptions, noting that high levels of immigration (legal or illegal) are often linked to economic prosperity. For example, undocumented workers enable the production of affordable consumer goods, partly because they accept terms of employment deemed unacceptable by legal residents. It also bears noting that illegal immigrants often pay taxes but do not use the public services to which they are entitled (either out of fear or because they lack required documents). When illegal immigration drops, furthermore, American workers do not rush to fill the jobs that theoretically become available. On the contrary, employers often wind up pressuring politicians to loosen immigration laws so that they can obtain needed labor. And immigration proponents argue that immigrants, legal or illegal, benefit the American economy because they are hardworking, motivated, and enterprising (Broder 2006; Lowenstein 2006).

It is difficult to assess the extent to which illegal immigration depresses wages because evidence is contradictory on this point. It is possible that employers do not raise wages because the supply of cheap illegal migrants exists. Alternatively, it is possible that these are simply low-wage jobs that would go unfilled if migrants stopped coming. Some support for the first argument can be found in the reality that it would not be all that difficult, economically, to raise wages in sectors where illegal immigrants work. For example, wages represent a tiny percentage of the cost of the fruits and vegetables that Americans consume. Raising wages significantly in that sector would result in a small (2% or 3%) increase in the cost of produce; thus, the argument that employers *can't* raise wages rings somewhat hollow (Broder 2006).

What should be done about illegal immigration? Most of the discussion has focused on stricter border enforcement, an approach that has, thus far, proven to be ineffective. Arresting and deporting large numbers of illegal immigrants also has been discussed, but the evidence suggests that many deportees simply return. Moreover, many illegal immigrants have children who were born in the United States, creating a moral dilemma for those seeking to deport the parents. Such children receive citizenship under the provision of the 14th Amendment, but some conservative legislators have called for a change in this rule because they accuse undocumented workers of deliberately having children (so-called anchor babies) to discourage deportation. The alternative is to focus on employers by strictly enforcing sanctions against those who hire illegal immigrants. This would have the effect of reducing demand for illegal immigrants (Cornelius 2005). However, employers and their advocates typically object to such policies, arguing that they unfairly burden them and that it is difficult to determine the legal status of prospective employees.[4] Additionally, as we discussed previously in this book, policy needs to respond not only to "pull factors" (what is attracting these workers to the work in the United States), but also the "push factors" (labor problems in the developing world and how poor job prospects at home compel workers in these societies to seek work elsewhere). A domestic work policy, especially as it relates to the issue of illegal immigration, would have to be coupled with an international work policy that ensures opportunity for workers irrespective of their nationalities (Kochan 2005).

Conclusion

Race, ethnicity, and work are commonly thought of as issues of the past, especially by those who are not directly or indirectly discriminated against. In this chapter, we show that the opportunities to work, and to succeed at work, remain strongly tied to issues of race and ethnicity, both because old problems persist and because new ones have emerged. They influence opportunities to obtain resources, to develop the skills needed to get good jobs, to tap into social networks that open doors, and to fit into work environments dominated by white and male power holders. Race and ethnicity also influence the opportunity to find jobs, which are often far removed from ethnic communities.

To shape a work policy for the new economy, one that is truly inclusive of all groups, race and ethnicity need to be included as important considerations. Part of the solution will necessarily involve creating greater equality of opportunity and directing more resources (e.g., finances, skills, and social connections) to minority groups that suffer greater poverties. This will require increased allocation of funds to underprivileged families and communities. The persistence of racial and ethnic preferences in hiring and the continued application of racial and ethnic stereotypes point to the need for further development of multicultural sensitivities. This will be challenging to accomplish as long as geographies divide the population into polarized ethnic enclaves. And efforts need to be made to promote the creation of good jobs in places where quality employment has largely disappeared. But perhaps some of the most effective initiatives to reduce racial inequalities will involve creating a situation in which those at the bottom of the economy can be adequately rewarded in this affluent society. This can be done by increasing the resources that come from jobs (e.g., by raising the minimum wage to a living wage) and by providing collective resources, irrespective of racial identity, class, and geographic location. This would help poor people of *all* racial and ethnic backgrounds, and would do much to promote racial equality as well.

There has been remarkable progress in advancing racial inclusion at work, but old interracial tensions remain and new ones are emerging. Far from having eliminated the problem, the new economy remains divided along racial and ethnic lines, and there are few indications that the most recent economic transformations are drawing these concerns to an end.

Notes

1. The U.S. Census treats *Hispanic* as a broad ethnic category that includes many different groups (Mexican Americans, Cuban Americans, Puerto Ricans, and so on). Some have objected to this term, both because it is artificial (clustering together groups who think of themselves as different) and because it includes groups who do not speak Spanish (for example, many Guatemalans or Peruvians speak native languages rather than Spanish). The term *Latino* has been suggested as an alternative, yet there is no consensus about which term is preferable. We have used

Latino primarily to emphasize the issue of ethnic diversities. However, in the tables, we have used the term *Hispanic,* as it is used in the government sources from which the data originate.

2. Because costs of living change over time, we converted these income figures to "constant dollars"—the values of these earnings are adjusted for the effects of inflation.

3. It should be added that these data consider only people who have incomes. Because African Americans are much less likely to be employed than either whites or Latinos, these data probably understate the gap in income between the average African American and the average member of other groups in American society.

4. For example, the U.S. Chamber of Commerce opposed provisions in the recent Arizona Immigration Law that made businesses responsible for enforcing immigration policies and imposed sanctions on those not in compliance. See: http://topics.law.cornell.edu/supct/cert/09-115 for a summary of their arguments.

Chapter Eight

Reshaping the Contours
of the New Economy

There is little evidence that the new economy is moving in a direction that will ensure everyone will have opportunities to engage in meaningful work, will earn a comfortable income, or will have the resources to construct satisfactory lives outside of work. On the contrary, employer practices and institutional arrangements continue to sustain—and sometimes even deepen—the chasms that separate workers from opportunity.

In this final chapter, we consider ways to reshape the new economy so that opportunities and resources are more equitably distributed. To do this, we first revisit the issue of opportunity divides to highlight the major concerns facing workers and discuss what would need to change to address those concerns. We then turn our attention to the process of social change and strategies to make workers' interests a priority in the ongoing development of the new economy and some obstacles that will be encountered along the way. Our hope is that strategic action, built through individual and collective efforts, can bring about needed changes. Finally, we conclude this book with a brief reflection on the prospects for liberated work in the new economy, work that would match the potential contributions—and needs—of a diverse workforce.

Opportunity Chasms

Class Chasms

One of the major problems in the new economy is that so many workers find it difficult or impossible to find satisfying and secure jobs that pay.

There are both old and new aspects to this problem. Although workers across the class spectrum face problems of strained schedules, insecure jobs, and uncertain futures, these concerns are felt most strongly by those working in the numerous "McJobs" that emerged as a result of efforts to deskill work in the old economy. Today, legions of workers labor in jobs that offer low pay, few benefits, and few opportunities for growth. These workers can expect scant rewards from their jobs and slim prospects that diligent efforts will result in upward mobility. And because women and minorities tend to be funneled into these low-end jobs, this class divide contributes to gender and racial inequalities.

While bad jobs continue to exist at the bottom of the American economy, there also have been efforts to chip away at the securities and leisure time available to those in the middle class. Insecurity extends to nearly all segments of the workforce, and most workers contend with strong prospects that their (or their spouse's) employer will close shop, lay them off, or restructure their jobs away. Their careers commonly require working long hours and affect their ability to form rewarding family lives. Access to these jobs increasingly depends on (but is not guaranteed by) access to education, and that, in turn, depends on whether one comes from a family with the resources to underwrite the costs of staying in school into early adulthood. These workers' lives remain very different from the lives of those laboring at the bottom, but a feature of the new economy is that many "good" jobs have acquired features once associated with less desirable employment. At the same time, the transition to the new economy witnessed the richest members of society becoming even wealthier and further separating their lives from the rest of society. In contemporary America, the significant gaps are not just between the middle class and the poor, but also between the affluent and everyone else. It is hard not to conclude that the good fortunes of those at the top of the class structure have come from the bad fortunes of those at the bottom and the increasingly fragile middle.

What can be done? One of the most pressing concerns is to address the economic problems faced by low-wage workers and to reshape the terms under which their jobs operate. There needs to be a national dialogue, and action, to rebalance the equation of a fair day's effort and a fair day's pay, including issues of compensation, scheduling, and security. But beyond those issues, resources and institutional arrangements also will need to be restructured in ways that enable workers to move from dead-end, low-skill positions into jobs that offer greater rewards. This will require, among other things, designing and supporting educational opportunities that fit the structure of the new economy. And because the returns to work are so lopsidedly allocated, the lion's share of the financial burden of paying for these changes must be borne by the affluent.

Issues of job insecurity, unmanageable schedules, limited (and sometimes nonexistent) vacations, summary dismissal, and a variety of other concerns cut across class lines. Creating entitlements to reasonable treatment and reinforced safety nets will help all workers, not just those who labor in bad jobs. Launching initiatives to address these concerns will require abandoning cultural mind-sets that divide groups of workers from one another. As long as low-wage workers see themselves as pitted against those with education, or the middle class sees itself as being pitted against those receiving welfare, the effort to mobilize support for collective resources will be hampered. As was the case in bringing about the Fair Labor Standards Act and Social Security legislation in the wake of the Great Depression, catalyzing change almost necessarily will require recruiting the fragile middle class into the struggle.

Gender Chasms

In the old economy, social policies, organizational cultures, job designs, and personal expectations were shaped by the assumption that men would be breadwinners and women would be homemakers. Within this gender regime, women's efforts in the home were defined as something other than real work, and care work was not given its economic due. In the new economy, women are nearly as likely as men to work outside of the home and to aspire to meaningful careers. And it is not simply women's desires that influenced their increased integration into the paid labor force—most families need two earners to make ends meet. Inequalities between men and women persist in no small degree because of the persistence of ideas and structures that place most of the burden of care work on women and impede their access to the most financially rewarding jobs.

Gendered work standards, modeled on what men used to bring to the workplace, are now structured into job designs and employers' definitions of who are "ideal workers" (Williams 2000). Workers have been socialized to accept a career mystique, an unsustainable set of expectations promoting an intense commitment to work (Moen and Roehling 2005). These structures and beliefs chafe against the resources available to the new workforce, one composed increasingly of dual-earner and single-parent families. How are workers responding? Most commonly by laboring long hours, shifting the timing of life events (such as marriage and childbirth), doing without, and shouldering the burden of increased stress. Their ability to manage also often is predicated on their ability to employ low-wage workers to care for children or aging parents. Recognizing these facts highlights the need to dismantle a gender regime that continues to assign men and women to different jobs inside and outside the home. But beyond countering the forces of socialization and interpersonal discrimination, structural changes will have to occur as well. Compensation will need to be recalibrated so that care work receives equitable returns as a form of labor. And career and job templates that assume workers have the capacity to work like men with full-time, stay-at-home spouses will need to be reconfigured.

Workers and employers in the new economy need to redefine how much work employees should be expected to perform and to redesign jobs to be reasonably compatible with what workers can provide across their careers. Efforts to develop a "one-size-fits-all" approach to work and family strains are unlikely to succeed, as the potential contributions of most workers vary over the life course. A more fruitful direction for change involves creating new templates for flexible careers (Moen and Sweet 2004). These can include opportunities to scale back work hours or to take time-outs from the labor force. And individuals and couples will need to have the resources to effectively plan careers so that they can land on their feet when jobs are lost.

The issues surrounding gender and work are commonly cast as "women's problems," but the reality is that these are family problems that affect men, women, children, aging parents, and their communities. Accelerating the pace and extending the range of reform will require bringing men into the movement to humanize work by persuading them of the benefits to be reaped from reconfigured gender roles and resources. This is actually an easy argument to make. Which stressed worker would not want to see the establishment of reasonable work schedules, expanded vacations, and opportunities to secure time-outs from his job? If more men are convinced that they lose out with gendered divisions of labor—particularly in their access to time to spend with their children—the clamor for change will increase, and coalitions for change will strengthen. Mobilizing this support requires recasting workplace reform as more than a question of gender equity.

Racial and Ethnic Chasms

Though Civil Rights–era legislation imposed legal barriers against discrimination, racial and ethnic ties continue to influence access to jobs. Today, race and ethnicity play a role in determining where one lives, the resources families have to pass from one generation to the next, and the attitudes gatekeepers hold about individuals from different backgrounds. Like the gender chasm, the race chasm is maintained through social structures that differentially allocate resources and opportunities, as well as cultural orientations that shape expectations.

If the new economy is to become truly race-blind, it will need to address—but also go beyond—sources of interpersonal discrimination, such as the widespread use of stereotypes that label potential employees as suspect. The disadvantages minority group members face stem from being raised in families that lack the economic resources to afford college tuition, the social connections to facilitate entry into good jobs, and the cultural capital that provides the soft skills needed for success in the new economy. But family disadvantages and interpersonal discrimination are not the only source of racial inequities. They also result from community-level resource deficits. The forces that concentrate members of underprivileged racial and ethnic minority groups into disadvantaged neighborhoods deprive members of these groups of quality education, information about work and careers, and the types of social ties that link them to jobs.

Racial tensions continue to divide the U.S. workforce, with controversies over immigration, affirmative action programs, and welfare reform all exhibiting strong racial overtones. The dialogue about whether illegal immigrants are "taking jobs from American workers" or "doing jobs that no one else will perform" distracts from the larger question of how to regulate low-wage work. A crucial step in reorienting this debate will be to shift the focus away from blaming the victims of globalization and economic restructuring, and toward analyzing the conditions of work—at home and abroad. Only of late has there been renewed consideration of the exploitative conditions under which most low-wage work occurs or of the role of employers who provide jobs to undocumented workers. Building racial coalitions will be a great challenge because prejudice and discrimination are etched so deeply into the culture and structure of American society. Still, these barriers can be dismantled by groups that share common interests, and as this happens, prospects for opening opportunities for underrepresented minority groups will be enhanced.

International Chasms

The new economy's global reach is profoundly affecting work opportunities around the world. It is impossible to ignore the remarkable transformations occurring in China, India, and many other developing nations. Longer life expectancies, as well as access to medical care, educational opportunities, and consumer goods, reflect a variety of positive outcomes of globalization. But at the same time, globalization has contributed to a number of new problems in the developing world, including pollution, overpopulation of cities, and the creation of hazardous work environments. Inequalities persist, both within and across developing nations.

The emergence of an affluent, educated business and professional class in countries such as China or Brazil has occurred in tandem with both old and new forms of low-income work in factories, in the informal economy, and the new cityscapes. Some societies are advancing, but other places in the global economy, especially in sub-Saharan Africa, have been largely left behind. The flow of work opportunities into the developing world and the movement of employment away from its traditional locations in the United States and Europe have contributed to American workers' feelings of insecurity and fueled international antagonisms.

The solutions to these problems are unlikely to come from the free market. As Karl Marx observed, the forces that make capitalism so productive are intertwined with the drive to exploit labor, and to do so around the globe:

> The need of a constantly expanding market for its products chases the bourgeoisie over the whole surface of the globe. It must nestle everywhere, settle everywhere, establish connections everywhere. . . . All old-established industries have been destroyed or are daily being destroyed. They are dislodged by new industries, whose introduction becomes a life and death question for all civilized nations, by industries that no longer work up indigenous raw material, but raw material drawn from the remotest zones; industries whose products are consumed, not only at home, but in every quarter of the globe. In place of the old wants, satisfied by the production of the country, we find new wants, requiring for their satisfaction the products of distant lands and climes. (Marx & Engels 1972 [1848])

The current system enables unfettered mobility of capital and creates incentives for employers to move jobs to locations where labor and the environment can be most easily exploited. Developing societies will not necessarily benefit from this movement, because the same mechanisms that enable capital to move also make it possible to shift the profits to other locations. Noted analysts such as Thomas Friedman (2005) optimistically assert that globalization will "flatten" the divides between developed and developing countries. Though this may be true in cultural terms, in economic terms the divides between developed and many developing countries are expanding (Milanovic 2005). This is not to say that life will not continue to improve in most developing nations, but it will not be nearly as good as it could be if reasonable checks and controls, designed to ensure more equitable returns from work, were placed on free-market capitalism. This needs to be done not only for ethical reasons, but also because it is in the interests of workers in the United States to do so (Kochan 2005).

The Agents of Change

Throughout this book, we posed questions concerning the contours of the new economy and the origins of the divides that separate workers from opportunity. In this final section, we consider two remaining important issues: identifying the types of changes that are feasible and the forces that can make these transformations happen. Our argument is that refashioning the new economy will require the concerted efforts of multiple agents, ranging from individual consumers to activist groups, unions, employers, government, and international organizations. However, not every agent will carry the same level of force, and when their efforts are misapplied (or misrepresented), they can actually impede change rather than accelerate it.

The Role of Individuals

Efforts to reshape the workings of the new economy, and to remedy its problems, often encourage a "do-it-yourself" approach. But can the cumulative efforts of disconnected individuals lead to the types of changes needed to refashion substantially the contours of work and opportunity? Consider, for example, the advice manual *The Better World Handbook: From Good Intentions to Everyday Actions* (Jones et al. 2002). This book (like a number of others) suggests that the path to reform leads through self-reflection and conscious efforts to "do the right thing." Toward that end, consumers are advised to limit their trade with companies that are less than environmentally sensitive or labor friendly, as well as to treat coworkers and subordinates with the dignity that they deserve. Investors are encouraged to buy stocks in companies that have a positive track record on issues of public concern. And, like innumerable other self-help books, it offers advice on the best ways to "balance" work and family.

Though intuitively appealing, these do-it-yourself solutions are unlikely to create truly meaningful change by themselves. One problem with the focus on the isolated consumer is that individuals face the Herculean challenge of identifying those brands and companies that are "responsible." To do so requires not only understanding the actions of distributors, but also the various companies that are linked to them in vast global supply chains. And companies can have mixed histories—for example, they may be strong on labor concerns but much weaker on environmental concerns. Socially responsible consumers also experience considerable difficulty determining which products to buy because the information they receive is often confusing and misleading (Nestle 2003; Seidman 2007). And even if consumers did have access to the relevant information, options to purchase responsibly do not always exist. This is a classic example of the exit/voice dilemma: individual consumers may possibly influence policy through exit (by avoiding choices they reject) but lack voice; that is, they cannot *create* options that correspond to their preferences. Of equal concern is the observation that consumers are reluctant to purchase fair trade products that cost much more than standard goods. In practice, people tend to place higher premiums on issues of style and price than on ethical business practices (Iwanow, McEachern, and Jeffrey 2005; De Pelsmacker, Driesen, and Rayp 2005).

The expectation that individuals will provide solutions for collective problems has a long history in American culture. For conservatives, individualism reflects an American virtue, and they reference Tocqueville ([1836] 1969) in their observations of the ways their country's successes hinged on Americans' embrace of autonomy and volunteerism. They tout the merits of an "ownership society" that allocates as many resources as possible to individuals. Thus, rather than invest in a national health care system or Social Security, the individualistic approach is to offer health "savings accounts" and personally managed retirement portfolios. Nor is it only conservatives who place individualistic perspectives at the forefront of social policy; so do many liberals. This can be witnessed, for example, in efforts to end "welfare as a way of life" and to introduce strict limits on support through the Personal Responsibility and Work Opportunity Reconciliation Act. In all of these cases, reforms were presented as empowerment, but actually had the consequence of shifting the responsibility for managing risk to individuals. As Jared Bernstein (2006) argues, these programs rest on an underlying philosophy that says "you're on your own" and that individuals have no right to expect support or protection from hardship (see also Hacker 2006).

Numerous cultural critics have argued that individualism results in a diminished capacity to empathize and understand how one's fortunes (or misfortunes) are connected to larger social processes (Bellah et al. 1985; Mills 1959; Putnam 2000; Riesman 2001). Even when sympathy exists, notice how individual volunteers are assigned the task of ameliorating hardship, supplanting bolder initiatives to challenge oppressive structural arrangements. For example, consider two of the most prominent volunteer and semivolunteer groups in America today: Habitat for Humanity and Teach for America. To date, Habitat for Humanity International, with a nearly all-volunteer membership, has built or repaired approximately 400,000 homes in America and abroad since its inception. In 2009, Teach for America had 4,000 dedicated young teachers working in impoverished school districts. Without disparaging these accomplishments, it is important to place them in context. In 2009, in the United States alone, *44 million* people lived in poverty and *one in 10* children did not graduate from high school. As well-intentioned as volunteers are, they commonly do little (if anything) to eliminate structural barriers or establish a more equitable distribution of social resources.

In sum, individualistic solutions are hampered by three fundamental concerns. First, people may not have the inclination or the capacity to behave in a manner that will transform the new economy. Second, individualistic efforts generally leave untouched the underlying forces that shape the contours of work and opportunity. Third, as a cultural framework, individualism reinforces the shift of risks to individuals, rather than building on the strength of collectivities. This does not mean that individuals cannot make a difference by trying to be socially conscious or that these efforts cannot create good outcomes. But these efforts *alone* will not be sufficient to address the root sources of the problems or make a significant impact on the operations of the new economy (Bernstein 2006).

The Role of Activist Groups

If individuals will not be the answer, what about activist groups? These groups comprise individuals who band together to exert pressure on governments, employers, and consumers. Their goals are to use collective action, primarily directed at the local level, to influence employment and trade practices, not only at home in America, but also abroad in developing countries. Here, we see greater cause for optimism, but the impact of these efforts necessarily will depend on recruitment, alliances with other organizations, and the ability to identify and target institutions that can effect change.

Consider the successes of activist groups, such as United Students Against Sweatshops and United Students for Fair Trade, which have influenced their own educational institutions to sever contracts with producers with histories of labor abuse (Crawford 2003; Glover 2003). Among their accomplishments is increased public awareness of exploitative labor practices and their links to the production of well-known consumer items, such as those endorsed by P. Diddy, Kathie Lee Gifford, and Michael Jordan. By putting pressure on institutions from within, and by mobilizing public opinion, these groups engage in a "name-and-shame" strategy to persuade employers to change their business practices (Seidman 2007).

Other grassroots organizations pressure local governments and employers to implement reasonable terms of employment. For example, activist groups such as Jobs with Justice operate almost entirely through local volunteers. Both of these groups organize social awareness campaigns and propose legislation at the local level to control employment practices (O'Brien and Gupta 2005). As a result of their efforts, more than 100 cities and jurisdictions had implemented living-wage ordinances by 2003 (Freeman 2005).[1] We witnessed these effects first hand, as a small group of students at Ithaca College (where one of the authors teaches) staged a series of press releases, articles in the student newspaper, and small demonstrations, demanding that the college pay all of its workers a living wage. The students used the college's image (or brand) for leverage, as the college presented itself as being a socially conscious organization devoted to sustainability. The students revealed to the wider community that the college was not fully living up to its ideals, as the institution subcontracted its food services to the subcontractor Sodexo, which did not pay its employees a living wage. In response to the pressure exerted by the students, the college renegotiated the terms of its agreement with Sodexo, which in 2011 instituted new wage rates at the Ithaca College campus.

Although these types of successes provide reason for hope, it is important to assess them relative to the magnitude of the problems. Even the most successful grassroots initiatives tend to have limited reach in sparking change in employment practices in the new economy. In part, they are hampered by limited resources and fluctuating memberships. When they do effect change, gains can be fleeting, as companies remain free to move facilities to "employer-friendly" locations (Armbruster-Sandoval 2005). Another problem is related to the strategy of focusing change at the local level, which can sometimes be trumped by more macrolevel initiatives. For example, companies can influence the creation of legislation at the state level to restrict the right of local communities to set their own labor standards. And focusing on the local level tends to produce changes that are confined to small groups of workers in particular locations. Careful study of living-wage initiatives, for example, reveals that comparatively few low-wage workers have been affected (most studies estimate the number to be less than 100,000, a small fraction of today's workforce) (Freeman 2005).

Arguably, the biggest success of activist groups has not been in local reforms, but in increasing public awareness of labor and environmental abuses. In response, companies have become more conscious of their public image, which in turn has influenced some major employers to reform labor and trade practices. For example, once the widespread use of child labor gained public recognition, companies such as Nike, Reebok, and Adidas agreed to monitor their suppliers' employment practices. Coffee distributors such as Starbucks have signed agreements with "preferred suppliers," who have promised to pay better wages, not to employ child labor, and to have their operations monitored (Schrage 2004). Still, critics, such as the Organic Consumers Association, claim that Starbucks makes only limited use of "fair trade" coffee beans and that the company's commitment to economic justice is superficial. As we discuss later, employers have developed a variety of strategies to maintain the impression of being socially responsible, while often avoiding being truly responsible (Seidman 2007).

Exhibit 8.1   Student Protesters Outside a Gap Store in Manhattan

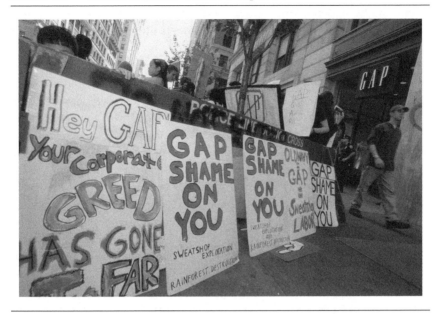

Source: Copyright © by Najlah Feanny/CORBIS SABA. Reprinted with permission.

The Role of Organized Labor

The American union movement has been in decline for much of the past 50 years and efforts to reverse this trend have proven largely unsuccessful. As we noted in Chapter 2, unions face a variety of problems in the new economy, including the decline of manufacturing employment, a legal framework that makes organizing difficult, capital mobility, the proliferation of smaller workplaces and subcontractors, and many more. The problems facing organized labor are compounded by public relations concerns. Many American workers have ambivalent feelings toward unions, which are commonly perceived as the reason why jobs move overseas, as protecting deadbeat employees, as corrupt, and as undermining the principle that individual workers (rather than classes of workers) should be compensated according to their efforts (Lichtenstein 2002). Only one in five Americans (20%) thinks that unions are "excellent" or "very good" for the country, about one in three (36%) believes that unions block economic progress, and two in three (69%) think unions have enough or too much power.[2]

Though unions have diminished power in the new economy, it is important to recognize that they still hold considerable sway. Unions play a vital role in influencing job contracts for those who are members or who work in unionized workplaces. Today, even though smaller proportions of the labor force are organized, they remain a substantial political force, particularly within the Democratic Party (Dark 2001). Approximately one in 10 workers is a union member, and many more Americans belong to a family in which at least one person is a union member.

There are some indications that unions could make a comeback in the United States. Despite ambivalent feelings about unions and their role in society, an estimated 40 million unorganized workers would vote for a union in their workplace if they were to be presented with the opportunity (Clawson 2003). The structure of the new economy and laws regulating union formation make organizing difficult, but organized labor does have new weapons at its disposal. For example, "just-in-time" production techniques leave little margin for error if supply chains are to be effectively coordinated. This creates the possibility for strategically managed strikes to disrupt an entire company's operations (Parker and Slaughter 1988). New patterns of labor migration also have brought to the United States new groups of workers. Some of these recent immigrants are difficult to organize: some are hostile to unions, others may have been frightened by labor repression at home, some are undocumented and fear deportation, and still others are "sojourners" who intend to return to their home countries. However, as was true in the 19th and 20th centuries, many immigrants (particularly those from Latin America) bring with them receptive attitudes to unionization and, in some cases, experience with organizing (Gordon 2005; Milkman 2006; Waldinger and Der-Martirosian 2000).

Several recent examples of successful organizing campaigns give proponents hope that the resurgence of union activity has already begun. In some cases, conventional unions have succeeded in organizing previously unorganized groups of workers. The Justice for Janitors campaign has made remarkable progress in organizing and winning better conditions for building maintenance workers (Milkman and Voss 2004). Similarly, the Hotel and Restaurant Workers Union succeeded in negotiating labor contracts with most of the full-service hotels in San Francisco (Wells 2000). Unions are even attracting the interest of professional workers, such as graduate students, who successfully gained union representation at several major universities (Lafer 2003). Surprisingly, some of the most successful organizing campaigns have been conducted by some of the oldest (and allegedly most "conservative") unions, those once affiliated with the American Federation of Labor (AFL). Ruth Milkman (2006) argues that this is because they are concentrated in growing sectors of the economy, in industries unaffected by capital mobility (such as hotels), and because their approaches are better suited to a volatile economy.[3]

Although conventional unions have had some successes, organizing low-wage workers, especially immigrants, sometimes requires another approach. One strategy that has shown promise is the creation of worker centers. These are community-based organizations, typically representing immigrants, that attempt to help low-wage workers in a variety of ways. They provide services such as legal help and education, they advocate, and they help to organize workers (Fine 2006). Some have succeeded in improving wages, stimulating organization, and making legislative gains for workers, even in places like suburban Long Island, where low-wage workers are often undocumented and isolated from one another in small workplaces (Gordon 2005).

Several future directions for the union movement have been proposed. One view argues that unions must adjust to changing economies and production practices (Heckscher 1988). In a system where jobs are constantly in flux, and where manufacturing only constitutes a small portion of the job market, the old-economy approach of using seniority systems to protect workers makes little sense. New methods for protecting workers are needed, some of which may be learned from earlier efforts. For example, rather than employers controlling who gets jobs, unions could collectively manage the placement of employees as opportunities expand and contract. This method of managing insecurity was commonly used in the craft union halls of the early 20th century and its time may have come again (Lichtenstein 2002). Others suggest that the future of unions depends on their regaining their militancy and energy. Only in this way, they argue, can unions overcome the negative image they have acquired and persuade workers that they are not bureaucratic, corrupt, and irrelevant (Lopez 2003). Dan Clawson (2003), for example, calls for a new "upsurge" of union activity, which would require unions to organize new workplaces and industries, and, most importantly, change the political and cultural climate in which unions exist. He argues that something like this happened in the 1930s, when the last big flurry of new organizing activity took place. He believes this can happen again, if unions are creative and respond to new problems and opportunities.[4]

A resurgent union movement will have to confront globalization and develop a means of collaborating and organizing workers on an international level. This would not be an altogether new idea. Even in the late 19th century, labor "internationals" existed, and many 20th century unions belonged to a variety of international union federations. More recently, an attempt to consolidate these international organizations led to the formation in 2006 of the International Trade Union Confederation (ITUC), which claims more than 300 affiliates in a wide range of countries. However, most international labor organizations, including the ITUC, do not engage in organizing efforts, but rather serve as clearinghouses for information, protest various forms of injustice against workers around the world, and pressure national and international organizations (such as the World Bank) on behalf of workers. Today, there are other opportunities for cross-national cooperation, such as an American union working together with a sister union at a branch plant overseas. Alternately, workers in one country could be organized to act in sympathy with workers in the same industry overseas. This happened in 1998, when American workers refused to unload ships affected by an Australian dock strike.

Unions will be better positioned to promote change when they integrate themselves with other activist groups, such as the antisweatshop campaigns and living-wage efforts mentioned earlier. In the past, unions and new social movements have not been receptive to one another. During the Vietnam War era, unions were often perceived (sometimes accurately) as opposed to the student movement. Nevertheless, when unions work in concert with other grassroots organizations, an approach that has been called *social movement unionism,* the results are often superior to when they operate alone (Gordon 2005; Milkman 2006). Successful living-wage campaigns illustrate this well; so do the much-discussed demonstrations in Seattle in 2001, in which labor groups cooperated with church groups, environmentalists, and others to protest the meetings of the World Trade Organization and its policies regarding global trade (Clawson 2003). And the remarkably successful organizing drives in Las Vegas hotels and casinos also owe much to a conscious strategy of working together with other community organizations (Fantasia and Voss 2004).

In short, labor organization could, again, become a powerful force for workplace reform. However, there is no guarantee that this will happen. Obstacles to union organization remain powerful and the gains discussed here are fragile. The New York University (NYU) graduate students' union, which organized in 2002 and made substantial gains in wages, benefits, and rights to overtime, was essentially shattered in 2005 (Arenson 2005; Epstein 2005). A much talked about effort to establish a union at one of the big-box retailers, Target, recently failed, illustrating the continued effectiveness of employer efforts to keep unions out (Greenhouse 2011b). Government action also can have a negative effect on unions' chances. There has been increased discussion of passing right-to-work laws in numerous states. These laws prohibit making union membership or the payment of union dues a condition for employment, and the passage of these types of laws will hamper both membership and the resources available to unions (Maher 2011). As we saw in Chapter 2, several major states have passed legislation that would, if not overturned, significantly limit public sector unions. Worker centers, too, often find their gains to be temporary or unenforceable and struggle both to forge links with nonimmigrant groups and to grow beyond a few hundred members (Gordon 2005). So, although it may be premature to announce the death of American unions, unions are unlikely to become suddenly powerful enough to reshape American workplaces by themselves.

The Role of Employers

The tensions between work and family life, and the growing awareness of the problems of global trade, are creating pressures to implement more socially responsible designs for workplace operations. But are these pressures sufficient to motivate employers to advance meaningful change from within? What is the likelihood that "internal" actions by the leading agents in the new economy will foster new approaches to work that serve the interests of workers and the communities in which they live?

One strategy to make employers take ownership of change relies on demonstrating how responsive policies can actually benefit companies. Work-family scholars call this the *dual agenda,* and posit that what is good for working families, or the wider society, does not necessarily come at a cost to employers. Some family-responsive policies, such as flexible work arrangements (when applied in specific types of settings to specific types of workers) can actually work in the employer's interests (Pavalko and Henderson 2006). Fulfilling the dual agenda requires demonstrating the potential for positive "returns on investment" that can be measured by increased productivity, greater profits, better employee retention, consumer loyalty, or other indicators of success. This **business case** for reform shows employers how their companies can benefit from altering production practices and taken-for-granted ways of working (Kossek and Fried 2006).

In many industries, a business case can be made to enhance labor standards through increased wages, benefits, or the creation of flexible work arrangements. For instance, recognition as a "best employer" in publications such as *Working Mother* magazine can help a company market itself to consumers and attract the most desirable employees. Similarly, a business case can be made to respond to brand tarnishing that results from naming-and-shaming pressures brought to bear by activist groups. A marketing business case can be made for developing products that fill the emerging niche for fair trade products. And opening opportunities for bridge jobs or flexible employment can provide a means to retain the talents, experience, and knowledge of an aging workforce (DeLong 2004). Researchers at the Massachusetts Institute of Technology (MIT) have worked with employers to design collaborative experiments to test how dismantling and replacing entrenched ways of organizing work affect the bottom line (Bailyn et al. 2006). Other researchers, in collaboration with employers, are documenting the extent to which companies are implementing worker-responsive practices and then publicizing findings from which companies can "benchmark" their own performance relative to that of the wider corporate community (Harrington and James 2006). Expanding the control workers have in managing their schedules has helped limit the extent of employee turnover and eased work-family conflict at companies such as Best Buy (Kelly, Moen, and Tranby 2011; Moen, Kelly, and Hill 2011).

Although these approaches show some promise, the employers in these studies tend to be those that want to retain and focus the energies of skilled workers. It is much harder, and sometimes impossible, to identify a business case to improve the conditions of work for those engaged in low-end jobs at home or abroad. If one traces the history of work from industrialization forward, it becomes apparent that the business case is often strongest for *dismantling*, rather than implementing, worker-responsive practices (such as providing high wages, secure jobs, health insurance, retirement benefits, and safe working environments—all of which have been eroded in the new economy). This is not to say that a dual agenda cannot sometimes be satisfied, especially for flexible work arrangements for highly skilled workers. But it is important to recognize the limitations of this approach because it is unlikely to close many of the major opportunity chasms, such as those that channel the lowest strata of the workforce into poorly paid, low-skill work. For those workers, the business case (established by history and numerous current cases) is to degrade work and opportunity.

Beyond the business case strategy, others suggest that employers may move the new economy forward as they embrace an ethos of **corporate citizenship** or **corporate social responsibility (CSR)**. In this case, the internal force is an open commitment, by corporate leadership, to understand their roles as stewards of the new economy (Blowfield 2005). CSR requires managers not to direct their actions simply to expand profit within the narrow confines of the law, but also to follow normative standards of professional conduct that go beyond what is obligatory or customary (Carter 2004). Perhaps most familiar to American consumers is Ben & Jerry's ice cream, a company that maintains that community-mindedness is a core component of the brand.[5] Beyond this company, the adoption of corporate social responsibility codes has been demonstrated to have a positive impact in the textile, leather tanning, and shoe industries in South Asia and in the wine industry in South Africa (Luken and Stares 2005; Nelson, Martin, and Ewert 2005; Winstanley, Clark, and Leeson 2002). And individual socially responsive leaders can make a difference in the lives of their workers. For example, because of the challenges of being a single parent following his wife's death, CEO Lewis Platt created sweeping changes at Hewlett-Packard, including the introduction of expansive flextime, flexplace, and family leave policies (Abelson 1999).

Will grassroots pressures and CSR lead to a more humane new economy? Gay Seidman (2007) offers a skeptical appraisal, basing her conclusions on the ways companies reformed (or failed to reform) their labor practices in South Africa, India, and Guatemala. She found that, even when the need for change was the focus of considerable public attention, corporate leaders largely failed to demonstrate a commitment to social responsibility. For example, even as the South African system of apartheid was exposed to widespread condemnation, American-based multinational corporations avoided challenging the South African government on human rights violations. Their primary goal was to keep operating within an oppressive regime. And when corporations did respond, they were slow, tempered, and acted in ways that generally respected the political system enforcing differential treatment of white and black South Africans. Even when companies signed pacts that required fair labor treatment, they were lax in enforcement, swayed evaluations to oversell their successes, and underreported the incidence and the implications of their failures.

Seidman concluded that companies, on their own, are reluctant to "do the right thing." Nor does pressure directly applied by consumers or activist organizations necessarily create a market-driven response that will lead corporations to "come to their senses" and act in responsible ways. A case in point is the much-publicized Rugmark program in India, a voluntary program in which manufacturers vow to not employ children in the manufacture of carpets. In return, participating employers receive a Rugmark tag that is sewn to the back of the carpet. This is then used as a marketing tool, giving consumers comfort in believing that the rug they are purchasing is made in accordance with fair labor practices. The Rugmark approach relies on a voluntary program that employers sign on to, which in turn is used to ratchet up the value of their product (and puts pressure on other employers to behave in a similar socially responsible fashion). Although Rugmark was initially lauded as an innovative market-based approach to stamping out child labor, subsequent analysis of its effects revealed serious flaws. One problem is lax oversight because the program is not supervised by the Indian government or international agents, but by a small number of inspectors employed by the Rugmark program. Inspections are rare, so that even when a rug is certified, it is uncertain that children were not involved. Rugmark also spawned competing marks, such as Care and Fair, that assure consumers that a portion of the proceeds from their purchase will be returned to benefit children and others in the carpet district. This latter approach leaves intact the practice of employing children in grueling, hazardous work, and leaves consumers with the false impression that their purchase is child labor–free. Another concern is that the voluntary approach to enforcing fair labor practices can foster a two-tiered distribution system, in which some products produced by the same manufacturer can be exported to societies (or stores) that care about fair labor practices, whereas others can be exported to societies that do not. Finally, as Seidman (2007) notes, fair trade labels may matter for luxury or visible product lines such as rugs, but they cannot be effectively applied to other products (such as ball bearings) that are hidden from the consumer.

An interesting, and potentially important, variation on the theme of corporate responsibility is the emergence of groups of activists and socially responsible businesses attempting to pioneer a new approach to economic activity. These groups, some of which are for-profit organizations, some of which are not, are developing alternatives to conventional ways of doing business, including implementing environmentally friendly, sustainable practices, creating jobs in poor urban communities abandoned by conventional businesses (e.g., the "green" businesses developed by the Evergreen Cooperatives in Cleveland), and experimenting with reward structures that limit income inequality (such as cooperative ownership or limits on the pay gap between CEOs and ordinary workers). The successes of some of these enterprises may inspire others to start new ones or may inspire other organizations to "rethink business as usual." On the other hand, pursuing these strategies does require a major change in conventional approaches to business activity, and other groups in society may find reasons to oppose them (for example, it is easy to present sustainability as expensive and hostile to job creation, especially in bad economic times) (Alperovitz 2011).

Is corporate leadership likely to guide companies to be "good citizens" in the new economy? Or are they more likely to continue to engage in "a race to the bottom" and seek ways to maximize profit at the expense of workers and the communities in which they live? As we suggest throughout this book, considerable evidence indicates that the latter is more likely. It is important to remember that capitalism itself discourages social responsibility because there are clear financial motivations to employ populations that can be most easily exploited and to move to locations where environmental protections are weakest. The case of Walmart, the world's largest employer and retailer, amply demonstrates this proposition. In the wake of widespread condemnation of its employment and trade practices, Walmart has directed far greater attention to its strategies for managing public opinion than it has to reforming its labor practices, and its profits and power continue to increase (Goldberg 2007). However, these outcomes are not inevitable, nor are organizations completely immune to social pressures or emergent industry standards. In fact, research that focuses on the ways organizations respond to normative pressures reveals that sometimes companies shift their practices because they see others doing likewise (DiMaggio and Powell 1983; Kelly 2003). From an institutional perspective, one might think of individual companies as being like the homeowner who mows his lawn because he sees his neighbors doing likewise. The more that practices such as flexible work arrangements are viewed as "taken-for-granted" ways of doing work, the more new companies will incorporate these practices into their talent management strategies.

The Role of Government

Political conservatives commonly caution against the interference of "big government" and the impediments regulation creates for the functioning of the economy. This perspective on government's role in the regulation of the new economy has a number of problems. First, government is *always* involved in economic activity. As the economic historian Karl Polanyi (1944) argued, the market itself could not exist unless the government had been there to help create it and continued to provide the political, legal, and military framework within which it operates. Second, government has a long history of making trade possible, and shielding workers from abusive work conditions. Most of the protections workers have today are the result of government intervention; these include child labor laws, unemployment insurance, environmental regulations, social security, minimum wage standards, rights to overtime pay, workplace safety oversight, the right to unionize, prohibitions against discrimination, and the right to family medical leave (see the appendix for a regulatory timeline). Those who espouse laissez-faire capitalism often overlook these necessary "intrusions" into the economy (Bernstein 2006; Galbraith 1976).

In the preceding chapters, we argued that the U.S. government generally has been reluctant to interfere in the contests between workers and their employers regarding pay, schedules, security, and other conditions of work. And U.S. government efforts pale in comparison with the more active role that governments have played in Western Europe. We also observed that, during the 20th century, the U.S. government went from being virtually absent from the regulation of work to being much more involved. The directions government actions will take in the future, and its speed of response, are not foregone conclusions. As in the old economy, competing interest groups will sway legislation, and the process will be political, not inevitable. Ultimately, government action will hinge on the ability of individuals, activist groups, unions, and others to influence its operations.

What role should the American government play in the new economy? The types of action we advocate will require reenvisioning the regulation of work, in much the same way as happened during the New Deal. This reenvisioning should consider specific workplace regulations, as well as the collective responsibility to provide for the basic needs of citizens, irrespective of their attachment to the labor force. We suggest three types of policy initiatives. First, there must be a revisiting of the issue of what constitutes reasonable conditions for work, and a revision of the standards enacted under the Fair Labor Standards Act (and other legislation) to consider both the nature of jobs and the composition of the labor force in the new economy. Second, there needs to be a national discussion of the means by which affluence and opportunities can be more equitably distributed. Third, though much attention has been focused on the U.S. government's role in promoting the interests of employers and protecting jobs at home, there needs to be a forceful discussion of its role in regulating the terms under which global supply chains operate and the role of the United States in fostering *positive* development abroad.

A Fair Labor Standards Act for the New Economy

When the Fair Labor Standards Act was enacted in 1938, it offered Americans, for the first time, the right to overtime pay and a minimum wage. Since its passage, its provisions have received modest updates and some new worker protections have been instituted. However, current regulations largely fail to address the needs of many workers in the new economy. Today, employers can expect workers to labor for below poverty-level wages; employers can impose long, short, or unpredictable schedules that interfere with employees' family responsibilities; and they are not obligated to provide information sufficient to enable employees to plan lives and careers. Redefining what constitutes the minimum standards of fair treatment is long overdue.

In 2011, the federal minimum wage remained at a scant $7.25 per hour and had a real value that was lower than it was 30 years earlier. Teenagers are commonly considered to be the typical minimum wage workers, but the reality is that nearly one in two (47%) minimum wage workers is older than age 25, and two in five (39%) work full time (Haugen 2003; Shulman 2005). And because the minimum wage is not tied to any objective standard, or automatically adjusted to correspond with inflation, its level will continue to be subject to the vagaries of political will.

During the Great Depression, the problem of unemployment was exacerbated because those who did have jobs were expected to work long hours, which in turn deprived others of the opportunity to work. The Fair Labor Standards Act responded by implementing overtime provisions that penalized employers for working their employees beyond 40 hours per week. By mandating time-and-a-half compensation for overtime, the act stimulated a reduction in work hours, as well as dispersed work opportunity. But because full-time employees are typically the only ones who receive benefits, and because the costs of benefits like health care have increased substantially, the penalty for overworking employees has decreased in the new economy. And because of changing opportunity structures, many more employees in the new economy are exempt from overtime provisions, as they labor in salaried positions and within organizational cultures that treat long hours as desirable and normal.

Today, the 40-hour work week plus time-and-a-half pay for overtime is a taken-for-granted arrangement, but one that needs to change. Although the current threshold may have been a reasonable expectation for a husband/breadwinner–wife/homemaker arrangement, today many dual-earner couples work a combined 80 to 100 hours per week (and sometimes more), leaving them frazzled and exhausted. One approach to fixing this problem can be found in France, a country that reduced the threshold for overtime eligibility to 35 hours per week. The result of this legislation has been to increase work opportunity and to decrease the level of stress shouldered by working families. One study of the impact of the French law found that nearly two in three workers responded that it had made it easier for them to combine work life with their family life (Fagnani and Letablier 2004). If shortening the workweek turns out not to be an option for the United States, an alternative would be increasing the penalty for overwork-ing employees—perhaps from time-and-a-half to double-time compensation.

A Fair Labor Standards Act for the New Economy needs to address the issue of overwork in other ways as well. For example, to bring the country more in line with the treatment of workers in Europe, Americans would need the opportunity to take 4 to 5 weeks of paid vacation a year. Current labor standards need to go far beyond the 12 weeks of unpaid leave pro-vided by the Family and Medical Leave Act and at a minimum, the current standard needs to be modified to mandate paid leaves of absence (which in turn will make family leave available to all workers). The variety of family leave and family supports enacted in Europe in the past 30 years provide valuable lessons to inform efforts to reshape American family leave policy (see Gornick and Meyers 2003; Bergman and Hobson 2002; Kelly 2006; Pfau-Effinger 2004).

Finally, a Fair Labor Standards Act for the New Economy needs to address the issue of job insecurities. In part, adjustments of family leave and overtime provisions may help to reduce insecurity by discouraging under-employment and forced "voluntary" exits from the labor force. Again, lessons from Europe may help to craft this legislation, as well as anticipate its impact on worker performance. In France, for example, complex *licen-ciement* laws regulate the conditions under which employees can be fired. To terminate a worker, companies commonly have to demonstrate that the employee could not be retained, show that the company cannot afford to keep the job in existence, or prove that the employee is incompetent. These laws substantially shift the onus of responsibility back to employers to legitimate why a good worker is being let go. However, this solution is not without problems. For example, in France when employees underperform, rather than firing them outright (as is commonly done in America), employers sometimes pay them to disappear graciously, and even go so far as to keep problematic employees on the books, but not assign them important work (Smith 2006). Though this type of legislation could have the result of keeping greater numbers of "deadbeat" employees on the payroll, French workers are actually more productive (as measured by productivity per hour worked) than American employees are (Krugman 2005). Whether American culture could embrace such commitments remains to be seen, but the key to winning the debate will be a concerted effort to shift the discussion from employer rights to worker rights (Shaiken 1984), and from "you're on your own" to "we're in this together" (Bernstein 2006).

Expanding Equities in Affluence and Opportunity

Some would argue that the U.S. government should not be in the business of equalizing opportunity, but the fact is that it has been active in this capacity for some time. Were it not for government intervention, rural America would not have paved roads or be electrified, those living downstream or downwind would drink and breathe far stronger toxins, and the public airwaves would not exist. The passage of the Civil Rights Act, Equal Pay Act, and Americans With Disabilities Act, as well as landmark court decisions such as *Brown v. Board of Education,* all demonstrate commitments to equalize opportunities. And government is the source of a number of entitlements—resources that all individuals have access to, regardless of their attachment to the labor force. One such entitlement is public education, which is available to all citizens through the 12th grade. Social Security, often presented as an "insurance" to which individuals "contribute," is also an entitlement, in that it provides incomes to older citizens and the disabled through the tax revenues generated by the current generation of workers.

At the domestic level, the primary impediment to the provision of collective supports is not scarcity of available resources but, rather, the *distribution* of those resources. In 2008, the most affluent fifth of the population received *half* of the nation's collective income. In contrast, the poorest fifth of the population received only a 3% share (Statistical Abstract of the United States 2010). Most other developed nations have implemented systems to offer broad access to quality health care, child care, and educational opportunities. Why can Americans, who live in one of the world's most prosperous economies, not have the same resources?

Health Care. In the past, many Americans lacked access to health care, because their jobs did not provide it, because they could not afford to purchase it privately, and because they were not eligible for government supported programs that aided the elderly (Medicare) and the poor (Medicaid). Efforts to make health insurance more broadly available were unsuccessful until, in 2010, President Obama signed into law the Affordable Care Act, legislation that would require all Americans to obtain health insurance and, if eligible, receive subsidies to purchase insurance or be enrolled in Medicare or Medicaid. With this new law, among other new protections, insurance companies are prohibited from excluding individuals because of preexisting conditions or from capping payments. However, as we write, a concerted effort is under way to challenge this law through both legislation and through court appeals. While the new law represents a remarkable advance, time will tell if the new approach is given a chance to work, how effective it will be in distributing health care opportunity, and whether unintended outcomes result.

Child Care. Access to quality child care is uneven, expensive, and often inadequate (Children's Defense Fund 2005; Heymann 2000). Today, low-income workers commonly lack access to good child care arrangements, and the middle class feels the financial pinch as well. One possibility is to follow the model offered by Finland and France, which have publicly funded high-quality day care facilities to help working parents remain in their jobs (Pfau-Effinger 2004). Instituting such changes in America will require demonstrating that child care centers can offer care that is as good (or better) than that which is received in the home. Establishing these centers will require directing public resources to constructing buildings, hiring trained personnel, and compensating them at the professional-level wages that care work deserves.

Ironically, as the need to care for disadvantaged children has increased, welfare reform efforts in the United States have undermined the prospects that poor single mothers can nurture their children (Crouter and Booth 2004). If the poor are expected to work, then jobs should pay wages sufficient to support families, and workers should have access to the resources to ensure that children have reasonable care and supervision. Lacking those resources, or the opportunity or ability to work, they should have sufficient public support to enable a reasonable standard of living.

Education. International comparisons of educational attainment are commonly used to make an argument that American public schools are not equipping students with the skills for work in the global economy. The reality is more complex. Those students who are fortunate enough to attend the better schools score as well—and usually higher—than students in nearly all other countries. However, students in disadvantaged school districts perform at low levels. Remedying class and racial divides will involve increasing resources to underfunded schools and combating racial and economic segregation. Beyond this obvious concern, there needs to be a national discussion about how to make a quality college education accessible to all members of society, especially to those students who lack financial resources (Kozol 2006; Marshall and Tucker 1992).

The Role of International Organizations

The latter part of the 20th century witnessed the creation of international trade agreements covering multiple countries. The most famous example is the North American Free Trade Agreement (NAFTA), but there are others around the world, including the creation of the European Union (EU), which goes well beyond a simple trade agreement. These arrangements are fueled by the belief that free trade promotes economic growth and that allowing unlimited capital mobility and tariff-free exchanges will boost all the economies involved, create employment, and generate profits. Because of this motivation, most attempts to assess the impact of these arrangements have focused on job creation. Critics argue that agreements like NAFTA encourage employers to shift employment to low-wage areas, but advocates respond that low-end jobs leave the United States, and new managerial and professional jobs are created as the result of business growth (Bair and Gereffi 2003; Gereffi 1994). Sorting out precisely what the economic impact of NAFTA has been has proven to be tremendously difficult.

NAFTA was accompanied by multinational side agreements on labor and environmental issues, which were intended to provide protections for workers and for the environment in all of the countries signing the agreement. Advocates for organized labor and for environmental causes complain that these have proven largely ineffective (Tseming 2004). Indeed, the fear is considerable that free trade will lead to a global race to the bottom, with countries reducing standards to stay competitive. But it could have been, and could still be, otherwise. If the best labor practices from each country became the standard in all of the participating nations, and if the enforcement mechanisms built into agreements like NAFTA are used effectively, this would improve the working conditions of workers throughout the region. There is much discussion of this within the EU, and in some ways, this may be happening—for example, in the efforts to standardize the maximum number of hours of work. The EU also has used its influence to require improved labor standards in nonmember countries with which it makes bilateral agreements. In other words, international economic agreements *can* become mechanisms for raising labor standards on a global scale.

Conversely, an opposite strategy of withdrawing trade can be used to bring countries with poor labor and environmental records into line. Should the United States "interfere" with the domestic policies of other countries in such a manner? The answer is that it has openly (as well as covertly) done so for some time. Currently, the U.S. government prohibits American businesses from trading with Myanmar (Burma), North Korea, Iran, and Cuba over human rights issues. And although the exploitation of workers abroad appears external to U.S. interests, it is one of the chief factors that stimulates the migration of jobs from the United States to developing countries. Here is a nexus where activist groups and unions can play a crucial role in influencing discussion of the conduct of international trade relations.

Human rights abuses at work come in a variety of forms, but one of the most devastating is child labor. UNICEF estimates that 246 million children work in the global economy and that approximately half are employed in hazardous environments. In Benares, India, as many as 200,000 children work in virtual bondage for manufacturers producing saris and carpets (Davis 2006). In Bangladesh, two-thirds of children aged 12 to 16 labor on average 8 to 10 hours per day (Delap 2001). Throughout the developing world, children are employed in a wide variety of endeavors in the apparel, shoe, fishing, rug, and construction industries. They are also major contributors to informal economies, working as (among other occupations) prostitutes, servants, and firewood collectors.

Some lessons can be learned from considering why child labor declined in Europe and America but persists in the developing world.[6] In the mid-19th century, children composed as much as 60% of the workforces in the factories that emerged in the wake of the Industrial Revolution (Gratton and Moen 2004). In the United States, children contributed one-quarter to one-third of the typical household's income in the early 20th century, which was more than the average contribution of wives (Cunningham 2000). Looking back at the decline of child labor in America reveals two forces at play. One force was cultural: children were redefined as needing protection (Aries 1965; Zelizer 1994). But equally and perhaps even more important were the changing rewards received from work, which made it possible for families to remove their children from the labor force and still maintain a reasonable quality of life (Cunningham 2000; Edmonds 2005). In developing nations, where per capita earnings are in the hundreds of dollars per year, a child's income can be critical to family survival. If standards of living rise in the developing world, we would anticipate child labor declines similar to those that occurred in the early 20th century in the United States. Reaching this goal almost certainly will require developing more effective means of regulating global supply chains.

One would hope that international agencies, such as the United Nations, would be well positioned to regulate global work, but this does not appear to be the case. Consider, for example, the United Nations' response to the problem of child labor in the new economy. In 1989, the United Nations ratified the following resolution:

> Children have the right to be protected from economic exploitation and from performing any work that is likely to be hazardous or to interfere with the child's education, or to be harmful to the child's health or physical, mental, spiritual, moral or social development. Additionally, states must (a) Provide for a minimum age or minimum ages for admission to employment; (b) Provide for appropriate regulation of the hours and conditions of employment; (c) Provide for appropriate penalties or other sanctions to ensure the effective enforcement of the present article. (United Nations 1989)

Although the United Nations condemns child labor, enforcement of this resolution has been hamstrung by vague standards on who is (or is not) a child, as well as how many hours and what conditions are "appropriate." Even in countries like India that have laws prohibiting the use of child labor, lax oversight of employers is common. As one analyst noted, it is ironic that India forbids child labor in the carpet weaving industry, yet the Indian government itself runs a training program for carpet weavers and recruits children as young as age 6. And though the world community condemns child labor, global investment practices (discussed below) continue to provide millions of dollars of funds to support the expansion of the carpet industry, with no barriers to prevent the use of child labor (Arat 2002).

The ability and the will of international organizations to go beyond merely symbolic gestures on the problem of labor abuse are questionable. The International Labor Organization (ILO) is the United Nations' branch concerned with labor questions. In principle, the ILO could become a force for improving the conditions of labor on a global scale and for using the United Nations enforcement authority to combat labor abuses in particular parts of the world. However, the ILO has largely defined itself as a clearinghouse for information, a center for research, and a forum for discussion of the problems of the workplace, but not as an enforcement agency.

Today, there appears to be a greater emphasis on opening trade than on regulating and controlling the terms under which it occurs. Consider, for example, the ways in which the World Bank and International Monetary Fund (IMF) approach the issue of fostering the development of the global economy. Although these are separate organizations formed for different reasons, by the end of the 20th century they shared three common concerns:

1. Crisis Management: to provide economic resources to help countries avert financial crises.
2. Transition: to foster countries' efforts to restructure to market-oriented economies.
3. Development: to help the poorest countries obtain resources to integrate themselves into global markets.

Their approach to resolving these problems has been to provide loans to countries in need, but under conditions that require political reforms in support of market-based economies. By providing loans to countries in need, the World Bank and IMF invest in high-risk economies, putting money where it is lacking. But the Structural Adjustment Programs that accompany these loans typically require reduced government supports for workers, privatization, diminished regulation, and a series of other measures that at best require no improvement in the conditions of labor and at worst lead to deterioration. These programs also provide little protection against loans being squandered through political waste and outright corruption. One of the IMF's and World Bank's goals is to foster economic development, but there is no clear evidence that they have been successful, and some evidence that their lending practices may have even undermined economic development. Such is the case of lending to sub-Saharan Africa. From 1971 to 1995, debts to the IMF and World Bank escalated from $10 billion to $230 billion. As a result, by 1995, the total debts owed by sub-Saharan African countries constituted nearly 70% of their gross national products, leading many societies to introduce austerity budgets and to cut back on needed investments in public education. Lacking the resources to pay these loans, these already poor countries are trapped in debt-borrowing spirals that now impoverish them, rather than help them develop (Babb 2005; Woods 2006).

The United States is in no position to tell other societies that they cannot use child labor, or to dictate wages and environmental practices. But it can dictate that, unless specific expectations are met, U.S. investors will be prohibited from engaging in trade with those societies. The first step is to initiate a national discussion of what these standards should be, and to foster a public awareness that it is in the interests of U.S. workers to have such standards enforced. Again, rather than focusing on influencing employers through a name-and-shame strategy, greater prospects for change may emerge from activist groups' efforts to pressure the U.S. government to support workers' movements at home and abroad (Seidman 2007).

The United States also could play a greater role in stimulating positive development. Despite being the major recipient of wealth generated in the global economy, the U.S. government lags far behind most countries in the share of its wealth it redistributes to developing countries. Of equal concern is the expectation that agencies such as the IMF and World Bank increasingly perform the needed development work. In 2010, the United States allotted less than 1% of its budget to foreign aid (only 18 cents per $100 of the country's GDP). The largest share of aid given was directed to military and political efforts and far smaller proportions were directed to true development efforts in impoverished regions of the world such as Africa and Southeast Asia. It also bears noting that the amount of aid being given, as a percentage of the United States' GDP, has declined steadily from the years following World War II to the present (Tarnoff and Nowels 2004).

Conclusion

In this book, we argued that the new economy has many new characteristics but that it also has integrated old practices into the design of jobs and the allocation of resources. Some of the problems evident today are new, but many reflect unresolved problems that emerged in the old economy, failed approaches to those problems, and lagging responses to emerging concerns. Today, America faces challenges analogous to those it confronted at the turn of the 20th century. Old means of working are being discarded, as new technologies and organizational practices are introduced. Previously dominant sectors of the economy are in decline, as new jobs and industries are emerging. And the workforce is changing, as are communities of workers. Along with the new challenges these changes present, there remain persistent concerns—of inequalities, overwork, unemployment, insecurity, and strained schedules.

If left unchanged, the contours of work in the new economy will be all too familiar. The future of work will be characterized by divided opportunities, with workers segregated on the basis of class, gender, race, and nationality. Some workers, particularly women and members of minority groups, will have careers dislodged or never make it onto career tracks at all. Large portions of the workforce will labor in alienating, low-skill, low-wage jobs. Some will have careers disrupted when they have children; others will keep their careers intact by foregoing having children altogether. Large numbers of workers will experience stress from having too much work, whereas many others will be work-poor and labor in jobs with unpredictable schedules, if they can find work at all. And most workers will labor under conditions of uncertainty and have inadequate resources to plan careers and to weather job loss.

Economies should work for people, not the other way around. The shape of the new economy is not immutable, nor is its future inevitable. One of the lessons learned by looking at work in historical perspective is that people have been able to effect positive change—even when faced with formidable resistance. Throughout this book, we present ideas for change and illustrations of how these changes have been introduced in other countries. A wide variety of existing examples show how societies and organizations have reduced overwork, fostered employment, expanded flexibilities, increased securities, and equalized opportunities.

Essential to solving all of the issues presented in this book is recognizing that the problem is not a scarcity of wealth or resources. Rather, the problem is the ways in which resources are allocated in a tremendously productive economy. There is no reason to accept the premise that good jobs and a good life inevitably will be beyond the reach of the hardworking men and women who make it all possible.

Notes

1. Such ordinances commit communities to do business only with employers who pay wages above minimum wage levels and that enable a family to live at a level above the federally defined poverty line.

2. Authors' analysis of the General Social Survey.

3. For example, Milkman notes that the AFL tries to get all (or most) of the employers in an area to agree to labor standards so that a single employer's threat to move or close down is less effective.

4. In 1995, when John Sweeney became leader of the AFL-CIO, he promised to rebuild the union movement by prioritizing organizing activities. The new regime did have some successes, most notably the Justice for Janitors campaign. However, the AFL-CIO's efforts in this regard may have stalled. Careful analysis of contemporary unions shows that only a few of the AFL-CIO's affiliates have made organizing new groups of workers a priority (Bronfenbrenner and Hickley 2003). In 2005, the AFL-CIO suffered a major split. A number of unions, including some of its largest and most vigorous members (SEIU, AFSCME) left to form their own federation (Change to Win) dedicated to more aggressive organizing. It remains to be seen whether this will divide and weaken the labor movement, or increase new organizing activity (Moberg 2005).

5. However, Ben & Jerry's has been purchased by an international conglomerate. There has been an effort to preserve the company's socially conscious image, but it remains to be seen whether this can be sustained over the long haul.

6. One notable contradiction in the United States is the continued acceptance of child labor on farms, where they are engaged in work that is dangerous, back-breaking, and largely tedious. Because farm workers are largely excluded from the protections of the Fair Labor Standards Act, children at age 12 (and in some instances even younger) can legally be employed for wages lower than the conventional minimum wage and with no limit on the number of hours the child may be compelled to work each day. Limited regulation of farm work and lax enforcement of the laws that do exist result in exposure to dangerous equipment and chemical contamination (Tucker 2000).

Appendix

Legislative and Regulatory Timeline of Worker Rights and Protections in the United States

Primarily adapted from Bureau of Labor Statistics (see http://www.bls .gov/opub/cwc/charts/cm20030124ar06c2.pdf) with information from other relevant sources included.

1884

Federal Labor Bureau, the predecessor of the Bureau of Labor Statistics, is established by the Hopkins Act.

1891

Kansas establishes the first state prevailing wage law.

1903

Department of Commerce and Labor is established by an act of Congress.

1912

Massachusetts adopts first minimum wage law for women and minors.

1913

U.S. Department of Labor is established by an act of Congress. It includes the Bureau of Labor Statistics, the Bureau of Immigration and Naturalization, and the Children's Bureau.

1914

Clayton Act limits the use of injunctions in labor disputes and provides that picketing and other union activities should not be considered unlawful.

1916

First federal child labor law is signed, but later struck down by the U.S. Supreme Court.

1920

Women's Bureau is established as a federal agency to represent the needs of wage-earning women in the public policy process.

1926

Railway Labor Act requires railroad employers to not discriminate against employees for joining a union and provides for the rights of workers to collectively bargain.

1931

Davis-Bacon Act provides for the payment of prevailing wage rates to laborers and mechanics employed by contractors and subcontractors on public construction.

1932

Norris-LaGuardia Act outlaws "yellow dog" contracts that required workers to sign away their rights to join unions as a precondition for employment.

1933

Wagner-Peyser Act creates the U.S. Employment Service in the Department of Labor.

1935

Federal Social Security Act provides a nationwide system of social insurance to protect wage earners and their families in old age, in the event of illness and disability, and in the event of economic hardship. Many key social programs of the 20th century developed from this act including Social Security, unemployment insurance, and Aid to Families With Dependent Children (AFDC). The National Labor Relation (Wagner) Act establishes the first national policy of protecting the rights of workers to organize and elect their representatives for collective bargaining purposes.

1936

Public Contracts (Walsh-Healy) Act sets labor standards on government contracts requiring the manufacture or purchase of materials.

1938

Fair Labor Standards Act sets minimum wage, maximum hours, and time pay, as well as equal pay and child labor standards.

1947

Labor-Management Relations (Taft-Hartley) Act reiterates policies protecting rights of workers to organize and elect union representatives. This act also places checks on union tactics, including secondary strikes and organizing procedures, and as such is commonly viewed by labor activists as a step backward in the labor movement.

1949

An amendment to the Fair Labor Standards Act directly prohibits child labor for the first time.
Courts decide that benefits are subject to collective bargaining.

1958

Welfare and Pension Disclosure Act requires administration of health insurance, pension, and supplementary unemployment compensation plans to file plan descriptions and annual financial reports with the Secretary of Labor.

1959

Labor-Management Reporting (Landrum-Griffin) Act prohibits improper activities by labor and management, such as secondary boycotts; provides certain protection for the rights of union members; and requires filing of certain financial reports by unions and employers.

1962

Manpower Development and Training Act requires the federal government to determine manpower requirements and resources and to "deal with the problems of unemployment resulting from automation and technological changes and other types of unemployment."

1963

Equal Pay Act prohibits wage differentials based on sex for workers covered by the Fair Labor Standards Act.

1964

Title VII of the Civil Rights Act establishes the U.S. Equal Employment Opportunity Commission to enforce federal statutes prohibiting employment discrimination.

1965

Medicare is established under Social Security. The McNamara-O'Hara Service Contract Act provides wage standards for employees performing work on federal service contracts.

1967

Age Discrimination in Employment Act makes it illegal to discharge, refuse to hire, or otherwise discriminate against persons ages 40 to 65.

1969

Federal Coal Mine Health and Safety Act protects the health and safety of the nation's coal miners.

1970

Occupational Safety and Health Act (OSHA) places certain duties on employers and employees to assure safe and healthful working conditions.

1974

Employer Retirement Income Security Act (ERISA) imposes standards on employer-provided benefit plans. The act is designed to protect the security of pension promises made by private sector firms.

1978

Pregnancy Discrimination Act requires employee benefit programs to treat pregnancy in the same way as illnesses.

1982

Job Training Partnership Act (JTPA) prepares youths and adults facing serious barriers to employment by providing job training and other services that would result in increased earnings, increased education and occupational skills, and decreased welfare dependency.

1985

Consolidated Omnibus Budget Reconciliation Act (COBRA) requires employers that provide health care benefits to continue such benefits to formerly covered individuals for a period of time after employer coverage ends.

1989

Worker Adjustment and Retraining Notification Act (WARN) provides protection to workers, their families, and their communities by requiring employers to provide notification 60 calendar days in advance of plant closings and mass layoffs.

1990

Americans With Disabilities Act (ADA) establishes a clear and comprehensive prohibition of discrimination on the basis of disability.

1993

Family and Medical Leave Act (FMLA) mandates employers to provide many types of workers up to 12 weeks of unpaid time off for worker and family medical purposes.

1996

The Personal Responsibility and Work Opportunity Reconciliation Act introduces work expectations for the receipt of welfare, replacing the earlier program of Aid to Families With Dependent Children (AFDC) with Temporary Assistance for Needy Families (TANF). Especially important changes are a maximum 5 years' cumulative support and the introduction of expectations that welfare recipients work (or prepare to work).

1998

The Workforce Investment Act (WIA) focuses on the needs of companies and how to make industries more productive by providing services that increase the number of jobs and the placement of individuals into those jobs.

1999

Ticket to Work and Work Incentives Act (TWWIA) provides health care and employment preparation and placement services to individuals with disabilities.

2009

Lilly Ledbetter Act amends the Civil Rights Act of 1964 to specify that the 180-day statute of limitations for filing an equal-pay lawsuit regarding pay discrimination resets with each new discriminatory paycheck.

2010

Fair Labor Standards Act is amended to require employers to provide reasonable break time and a private, nonbathroom place for nursing mothers to express breast milk during the workday, for one year after the child's birth.

References

Abelson, Reed. 1999. "A Push From the Top Shatters a Glass Ceiling." *New York Times,* August 22.

Alba, Richard, and Victor Nee. 2003. *Remaking the American Mainstream: Assimilation and Contemporary Immigration.* Cambridge, MA: Harvard University Press.

Alperovitz, Gar 2011. "The New-Economy Movement." *The Nation,* June 13.

Altonji, Joseph, and Jennifer Oldham. 2003. "Vacation Laws and Annual Work Hours." *Federal Reserve Bank of Chicago Economic Perspectives* 27(3):19–29.

Altucher, Kristine, and Lindy B. Williams. 2003. "Family Clocks: Timing Parenthood." *It's About Time: Couples and Careers,* edited by Phyllis Moen, 49–59. Ithaca: Cornell University Press.

Aman, Carolyn, and Paula England. 1997. "Comparable Worth: When Do Two Jobs Deserve the Same Pay?" In *Subtle Sexism: Current Practice and Prospects for Change,* edited by Nijole Benokraitis, 297–314. Thousand Oaks, CA: Sage.

American Association of Pediatrics. 2005. "Breastfeeding and the Use of Human Milk." *Pediatrics* 115:496–506.

American Psychological Association. 2011. *Stress in the Workplace.* Washington DC: American Psychological Association.

Amott, Teresa, and Julie Matthaei. 1996. *Race, Gender and Work: A Multicultural Economic History of Women in the United States.* Boston: South End Press.

Anderson, Elijah 1999. *Code of the Street: Decency, Violence and the Moral Life of the Inner City.* New York: Norton.

Appelbaum, Eileen, and Rosemary Batt. 1992. *The New American Workplace: Transforming Work Systems in the United States.* Ithaca, NY: ILR Press.

Aquilino, William S. 2005. "Impact of Family Structure on Parental Attitudes Toward the Economic Support of Adult Children Over the Transition to Adulthood." *Journal of Family Issues* 26:143–67.

Arat, Zehra F. 2002. "Analyzing Child Labor as a Human Rights Issue: Its Causes, Aggravating Policies, and Alternative Proposals." *Human Rights Quarterly* 24:177.

Archibold, Randall. 2010. "Arizona Enacts Stringent Law on Immigration." *New York Times,* April 24.

Arenson, Karen. 2005. "N.Y.U. Moves to Disband Graduate Students Union." *New York Times,* June 17.

Aries, Phillipe. 1965. *Centuries of Childhood: A Social History of Family Life*. New York: Vintage.

Armbruster-Sandoval, Ralph. 2005. "Workers of the World Unite? The Contemporary Anti-sweatshop Movement and the Struggle for Social Justice in the Americas." *Work and Occupations* 32:464–85.

Armstrong-Stassen, Marjorie. 1993. "Survivors' Reactions to a Workforce Reduction." *Revue Canadienne des Sciences de Administration* 10:334–44.

———. 1998. "The Effect of Gender and Organizational Level on How Survivors Appraise and Cope With Organizational Downsizing." *The Journal of Applied Behavioral Science* 34:125–42.

Arnett, Jeffrey. 2004. *Emerging Adulthood: The Winding Road From the Late Teens Through the Twenties*. New York: Oxford University Press.

———. 2007. "Afterword: Aging Out of Care: Toward Realizing the Possibilities of Emerging Adulthood." *New Directions for Youth Development* 113:151–62.

Arthur, Michael B., and Denise M. Rousseau. 1996. "Introduction: The Boundaryless Career as a New Employment Principle." In *The Boundaryless Career: A New Employment Principle for a New Organization Age*, edited by Michael B. Arthur and Denise M. Rousseau, 3–20. New York: Oxford.

Arthur, Michelle, and Alison Cook. 2004. "Taking Stock of Work-Family Initiatives: How Announcements of 'Family-Friendly' Human Resource Decisions Affect Shareholder Value." *Industrial and Labor Relations Review* 57:599–613.

Aspray, William. 2010. "IT Offshoring and American Labor." *American Behavioral Scientist* 53:962–82.

Astin, Helen S., and Jeffrey F. Milem. 1997. "The Status of Academic Couples in U.S. Institutions." In *Academic Couples: Problems and Promises*, edited by Marianne A. Ferber and Jane W. Loeb,128–55. Chicago: University of Illinois.

Autor, David, Lawrence Katz, and Melissa Kearney. 2006. "The Polarization of the U.S. Labor Market." *American Economic Association Review Papers and Proceedings* 96(2):189–94.

Babb, Sarah. 2005. "The Social Consequences of Structural Adjustment: Recent Evidence and Current Debates." *Annual Review of Sociology* 31:199–222.

Bailyn, Lotte, Ann Bookman, Mona Harrington, and Thomas Kochan. 2006. "Work-Family Interventions and Experiments: Workplaces, Communities, and Society." In *The Work and Family Handbook: Multi-disciplinary Perspectives, Methods, and Approaches*, edited by Marcie Pitt-Catsouphes, Ellen E. Kossek, and Stephen Sweet, 73–99. Mahwah, NJ: Lawrence Erlbaum.

Bair, Jennifer, and Gary Gereffi. 2003. "Upgrading, Uneven Development, and Jobs in the North American Apparel Industry." *Global Networks* 3:143–69.

Bandura, Albert. 1982a. "Self-Efficacy Mechanism in Human Agency." *American Psychologist* 37:122–47.

———. 1982b. "The Self and Mechanisms of Agency." In *Psychological Perspectives on the Self*, edited by Jerry Suls, 3–39. Hillsdale, NJ: Lawrence Erlbaum.

Barnett, Rosalind Chait, and Karen C. Gareis. 2000. "Reduced-Hours Employment: The Relationship Between Difficulty of Trade-Offs and Quality of Life." *Work and Occupations* 27:168–87.

Barrett, Rowena. 2004. "Working at Webboyz: An Analysis of Control Over the Software Development Labour Process." *Sociology* 38:777–94.

Bartlett, Donald, and James Steele. 1992. *America: What Went Wrong?* Kansas City, MO: Andrews McMeel.

Batt, Rosemary, Virginia Doellgast, Hyunji Kwon, Mudit Nopany, Priti Nopany, and Anil da Costa. 2005. *The Indian Call Centre Industry: National Benchmarking Report Strategy, HR Practices, and Performance.* Ithaca, NY: Cornell University Press.

Beck, Ulrich. 2000. *The Brave New World of Work.* Cambridge, UK: Polity Press.

Becker, Gary S. 1981. "Division of Labor in Households and Families." In *A Treatise on the Family*, edited by Gary S. Becker, 30–53. Cambridge, MA: Harvard University Press.

Becker, Penny Edgell, and Phyllis Moen. 1999. "Scaling Back: Dual-Earner Couples' Work-Family Strategies." *Journal of Marriage and the Family* 61:995–1007.

Beers, Thomas. 2000. "Flexible Schedules and Shift Work: Replacing the '9 to 5' Workday?" *Monthly Labor Review* 123:33–40.

Bell, Daniel. 1973. *The Coming of Post-industrial Society.* New York: Basic Books.

Bellah, Robert, Richard Madsen, William Sullivan, Ann Swidler, and Steven Tipton. 1985. *Habits of the Heart: Individualism and Commitment in American Life.* New York: Harper & Row.

Bem, Sandra. 1993. *The Lenses of Gender: Transforming the Debate on Gender Inequality.* New Haven, CT: Yale University Press.

Bennhold, Katrin. 2008. "Sarkozy Wins Battle to Loosen 35-Hour Workweek." *New York Times.* July 24.

Benokraitis, Nijole. 1997. "Sex Discrimination in the 21st Century." In *Subtle Sexism: Current Practice and Prospects for Change*, edited by N. Benokraitis, 5–33. Thousand Oaks, CA: Sage.

Bensman, David, and Roberta Lynch. 1987. *Rusted Dreams: Hard Times in a Steel Community.* Berkeley: University of California Press.

Bergman, Helena, and Barbara Hobson. 2002. "Compulsory Fatherhood: The Coding of Fatherhood in the Swedish Welfare State." In *Making Men Into Fathers: Men, Masculinities and the Social Politics of Fatherhood*, edited by Barbara Hobson, 92–124. Cambridge: Cambridge University Press.

Bernhardt, Annette, Martina Morris, Mark Handcock, and Marc Scott. 2001. *Divergent Paths: Economic Mobility in the New American Labor Market.* New York: Russell Sage Foundation.

Bernstein, Jared. 2006. *All Together Now: Common Sense for a Fair Economy.* San Francisco: Berrett-Koehler.

Bertrand, Marianne, and Sendhil Mullainathan. 2004. "Are Emily and Greg More Employable Than Lakisha and Jamal? A Field Experiment on Labor Market Discrimination." *The American Economic Review* 94:991–1013.

Bidwell, Matthew, and Forrest Briscoe. 2009. "Who Contracts? Determinants of the Decision to Work as an Independent Contractor Among Information Technology Workers." *Academy of Management Journal* 52:1148–68.

Blair-Loy, Mary. 2003. *Competing Devotions: Career and Family Among Women Executives*. Cambridge, MA: Harvard University Press.

Blau, Francine, and Lawrence Kahn. 2006. "The U.S. Gender Pay Gap in the 1990s: Slowing Convergence." *Industrial Relations Section Working Paper 508*.

Blinder, Alan S. 2006. "Offshoring: The Next Industrial Revolution?" *Foreign Affairs* 85:113–28.

Blowfield, Michael. 2005. "Corporate Social Responsibility: Reinventing the Meaning of Development?" *International Affairs* 81:515–24.

Bluestone, Barry, and Bennett Harrison. 1982. *The Deindustrialization of America: Plant Closings, Community Abandonment and the Dismantling of Basic Industry*. New York: Basic Books.

Bocian, Debbie Gruenstein, Wei Li, and Keith Ernst. 2010. *Foreclosures by Race and Ethnicity: The Demographics of a Crisis*. Washington, DC: Center for Responsible Lending.

Bookman, Ann. 2004. *Starting in Our Own Backyards: How Working Families Can Build Community and Survive the New Economy*. New York: Routledge.

Boris, Eileen, and Carolyn Lewis. 2006. "Caregiving and Wage-Earning: A Historical Perspective on Work and Family." In *The Work and Family Handbook: Multi-disciplinary Perspectives, Methods and Approaches*, edited by Marcie Pitt-Catsouphes, Ellen E. Kossek, and Stephen Sweet, 73–98. Mahwah, NJ: Lawrence Erlbaum.

Borjas, George, Richard Freeman, Lawrence Katz, John DiNardo, and John Abowd. 1997. "How Much Do Immigration and Free Trade Affect Labor Market Outcomes." *Brookings Papers on Economic Activity* 1997:1–90.

Bourdieu, Pierre. 1984. *Distinction: A Social Critique of the Judgment of Taste*. Cambridge, MA: Harvard University Press.

———. 1986. "The Forms of Capital." In *Handbook of Theory and Research for the Sociology of Education*, edited by John G. Richardson, 241–58. New York: Greenwood Press.

Bourgois, Phillippe. 1995. *In Search of Respect: Selling Crack in el Barrio* New York: Cambridge University Press.

Bowen, William, and Dereck Bok. 1998. *The Shape of a River: Long-Term Consequences of Considering Race in College and University Admissions*. Princeton, NJ: Princeton University Press.

Boydston, Jeanne. 1990. *Home and Work: Housework, Wages, and the Ideology of Labor in the Early Republic*. New York: Oxford University Press.

Bradsher, Keith. 2002. "A High-Tech Fix for One Corner of India." *New York Times*, December 27.

Branson, Douglas. 2010. *The Last Male Bastion*. New York: Routledge.

Braverman, Harry. 1974. *Labor and Monopoly Capital*. New York: Monthly Review Press.

Breen, T. H. 1985. *Tobacco Culture: The Mentality of the Great Tidewater Planters on the Eve of the Revolution*. Princeton, NJ: Princeton University Press.

Brockner, Joel. 1990. "Scope of Justice in the Workplace: How Survivors React to Co-worker Layoffs." *The Journal of Social Issues* 46:95–106.

Brockner, Joel, Steven Grover, Thomas Reed, Rocki Lee DeWitt, and Michael O'Malley. 1987. "Survivors' Reactions to Layoffs: We Get By With a Little Help From Our Friends." *Administrative Science Quarterly* 32:526–41.

Brockner, Joel, Batia M. Wiesenfeld, and Christopher L. Martin. 1995. "Decision Frame, Procedural Justice, and Survivors' Reactions to Job Layoffs." *Organizational Behavior and Human Decision Processes* 63:59–68.

Broder, John. 2006. "Immigrants and the Economics of Hard Work." *New York Times,* April 2.

Brody, Hugh. 2002. *Maps and Dreams.* New York: Pantheon.

Broman, Clifford L., V. Lee Hamilton, and William S. Hoffman. 1990. "Unemployment and Its Effects on Families: Evidence From a Plant Closing Study." *American Journal of Community Psychology* 18:643–59.

Bronfenbrenner, Kate, and Robert Hickley. 2003. "Changing to Organize: A National Assessment of Union Strategies." In *Rebuilding Labor: Organizers and Organizing in the New Labor Movement,* edited by Ruth Milkman and Kim Voss, 17–61. Ithaca, NY: ILR Press.

Brooks, David. 2005. "The Education Gap." *New York Times,* September 25.

Brooks, Michael. 2001. "Law Enforcement Physical Fitness Standards and Title VII." *FBI Law Enforcement Bulletin* 70:26–32.

Browne, Irene, and Ivy Kennelly. 1999. "Stereotypes and Realities: Images of Black Women in the Labor Market. In *Latinas and African American Women at Work: Race, Gender and Economic Inequality,* edited by Irene Browne, 302–26. New York: Russell Sage Foundation.

Buchanan, Patrrick. 2006. *State of Emergency: The Third World Invasion of America.* New York: Thomas Dunne Books.

Budig, Michelle. 2006. "Intersections on the Road to Self-Employment: Gender, Family and Occupational Class." *Social Forces* 84:2223–39.

Burawoy, Michael. 1979. *Manufacturing Consent: Changes in the Labor Process Under Monopoly Capitalism.* Chicago: University of Chicago Press.

Burke, Ronald J., and Esther Greenglass. 1999. "Work-Family Conflict, Spouse Support and Nursing Staff Well-Being During Organizational Restructuring." *Journal of Occupational Health Psychology* 4(4):327–36.

Buss, Terry F., and F. Stevens Redburn. 1987. "Plant Closings: Impacts and Responses." *Economic Development Quarterly* 1:170–77.

Butler, John Sibley, and Cedric Herring. 1991. "Ethnicity and Entrepreneurship in America: Toward an Explanation of Racial and Ethnic Group Variations in Self-Employment." *Sociological Perspectives* 34:79–94.

Callahan, Raymond. 1962. *Education and the Cult of Efficiency.* Chicago: University of Chicago Press.

Cappelli, Peter. 1997. *The New Deal at Work.* Cambridge, MA: Harvard Business School Press.

———. 1999. "Career Jobs Are Dead." *California Management Review* 42: 168–79.

———. 2008. *Talent on Demand: Managing Talent in an Age of Uncertainty.* Cambridge, MA: Harvard Business School Press.

Card, David. 1990. "The Impact of the Mariel Boatlift on the Miami Labor Market." *Industrial and Labor Relations Review* 43.

———. 2001. "Immigrant Inflows, Native Outflows, and the Local Labor Market Impacts of Higher Immigration." *Labor Economics* 19:22–64.

Carr, Deborah. 1996. "Two Paths to Self Employment? Women's and Men's Self-Employment in the United States, 1980." *Work and Occupations* 23:26–53.

Carter, Craig R. 2004. "Purchasing and Social Responsibility: A Replication and Extension." *Journal of Supply Chain Management: A Global Review of Purchasing & Supply* 40:4–16.

Castells, Manuel. 2000. *The Rise of the Network Society.* Oxford, UK: Blackwell.

Ceci, Stephen, Wendy Williams, and Susan Barnett. 2009. "Women's Underrepresentation in Science: Sociocultural and Biological Considerations." *Psychological Bulletin* 135:218–61.

Central Intelligence Agency, 2011. *The World Factbook.* Washington DC: Central Intelligence Agency.

Cha, Kyung-Wook, Robert Weagly, and Laura Reynolds. 2005. "Parental Borrowing for Dependent Children's Higher Education." *Journal of Family and Economic Issues* 26:299–321.

Chakravartty, Paula. 2001. "Flexible Citizens and the Internet: The Global Politics of Local High-Tech Development in India." *Emergences: Journal for the Study of Media & Composite Cultures* 11:69–88.

Chandler, Alfred. 1990. *Scale and Scope: The Dynamics of Industrial Capitalism.* Cambridge, MA: Belknap Press.

Chase-Dunn, Christopher, Yukio Kawano, and Benjamin D. Brewer. 2000. "Trade Globalization Since 1795: Waves of Integration in the World-System." *American Sociological Review* 65:77–95.

Children's Defense Fund. 2005. *State of America's Children 2005.* Washington, DC: Children's Defense Fund.

Chodorow, Nancy. 1978. *The Reproduction of Mothering.* Berkeley: University of California Press.

Chronicle of Higher Education. 2010. *Almanac of Higher Education 2010.* Washington DC: Chronicle of Higher Education.

Clawson, Dan. 2003. *The Next Upsurge: Labor and the New Social Movements.* Ithaca, NY: ILR Press.

Cohen, Barney. 2003. "Urban Growth in Developing Countries: A Review of Current Trends and a Caution Regarding Existing Forecasts." *World Development* 32:23–51.

Cohen, Ruth Schwartz. 1985. *More Work for Mother: The Ironies of Household Technology From the Open Hearth to the Microwave.* New York: Basic Books.

Collins, Sharon M. 1997. "Black Mobility in White Corporations: Up the Corporate Ladder but out on a Limb." *Social Problems* 44:55–67.

Committee on Maximizing the Potential of Women in Academic Science and Engineering, National Academy of Sciences, National Academy of Engineering, and Institute of Medicine. 2006. *Beyond Bias and Barriers: Fulfilling the*

Potential of Women in Academic Science and Engineering. Washington DC: National Academies Press.

Coontz, Stephanie. 1992. *The Way We Never Were: American Families and the Nostalgia Trap.* New York: Basic Books.

———. 1997. *The Way We Really Are: Coming to Terms With America's Changing Families.* New York: Basic Books.

Cooper, Michael, and Mary Williams Walsh. 2011. "Public Pensions, Once Off Limits, Face Budget Cuts." *New York Times,* April 25.

Cornelius, Wayne. 2005. "Controlling 'Unwanted' Immigration: Lessons From the United States, 1993–2004." *Journal of Ethnic and Migration Studies* 31:775–94.

Correll, Shelley. 2001. "Gender and the Career Choice Process: The Role of Biased Self-Assessments." *American Journal of Sociology* 106:1691–730.

———. 2004. "Constraints into Preferences: Gender, Status, and Emerging Career Aspirations." *American Sociological Review* 69:93–113.

———. 2007. "Getting a Job: Is There a Motherhood Penalty?" *American Journal of Sociology* 112:1297–1338.

Costa, Dora. 2000. "Hours of Work and the Fair Labor Standards Act: A Study of Retail and Wholesale Trade, 1938–1950." *Industrial and Labor Relations Review* 53:648–64.

Cottle, Thomas J. 2001. *Hardest Times: The Trauma of Long Term Unemployment.* Westport, CT: Praeger.

Cowie, Jefferson. 2001. *Capital Moves: RCA's 70-Year Quest for Cheap Labor.* New York: New Press.

Crawford, Elizabeth. 2003. "Good to the Last Drop." *Chronicle of Higher Education* 49:A8.

Creamer, Elizabeth, and Associates. 2001. *Working Equal: Academic Couples as Collaborators.* New York: RoutledgeFalmer.

Crittenden, Ann. 2001. *The Price of Motherhood.* New York: Henry Holt.

Crouter, Ann, and Alan Booth. 2004. *Work-Family Challenges for Low-Income Parents and Their Children.* Mahwah, NJ: Lawrence Erlbaum.

Cunningham, Hugh. 2000. "The Decline of Child Labour: Labour Markets and Family Economies in Europe and North America Since 1830." *Economic History Review* 53: 409–28.

Curry, James. 1993. "The Flexibility Fetish." *Capital and Class* 50:99–126.

Dark, Taylor. 2001. *The Unions and the Democrats: An Enduring Alliance.* Ithaca, NY: Cornell University Press.

Darrah, Charles. 2006. "Ethnography and Working Families." In *The Work and Family Handbook: Multi-disciplinary Perspectives, Methods, and Approaches,* edited by Marcie Pitt-Catsouphes, Ellen E. Kossek, and Stephen Sweet, 367–386. Mahwah, NJ: Lawrence Erlbaum.

Davis, Mike. 2006. *Planet of Slums.* London: Verso Books.

de Graaf, John, David Wann, and Thomas Naylor. (2001). *Affluenza: The All-Consuming Epidemic.* San Francisco: Berrett-Koehler.

Delap, Emily. 2001. "Economic and Cultural Forces in the Child Labour Debate: Evidence From Urban Bangladesh." *Journal of Development Studies* 37:1.

Dellinger, Kirsten, and Christine Williams. 2002. "The Locker Room and the Dorm Room: Workplace Norms and the Boundaries of Sexual Harassment in Magazine Editing." *Social Problems* 49:242–57.

DeLong, David. 2004. *Lost Knowledge: Confronting the Threat of an Aging Workforce*. New York: Oxford University Press.

De Pelsmacker, Patrick, Liesbeth Driesen, and Glenn Rayp. 2005. "Do Consumers Care About Ethics: Willingness to Pay for Fair Trade Coffee." *Journal of Consumer Affairs* 39:363–86.

Deutsch, Francine. 1999. *Halving It All: How Equally Shared Parenting Works*. Cambridge, MA: Harvard University Press.

Deyo, Frederic. 1989. *Beneath the Miracle: Labor Subordination in the New Asian Industrialisation*. Berkeley: University of California Press.

Dill, Bonnie Thorton. 1988. "Our Mothers' Grief: Racial-Ethnic Women and the Maintenance of Families." *Journal of Family History* 13:415–31.

DiMaggio, Paul, and W. W. Powell. 1983. "The Iron Cage Revisited: Institutional Isomorphism and Collective Rationality in Organizational Fields." *American Sociological Review* 48:147–60.

Dominguez, Silvia, and Celeste Watkins. 2003. "Creating Networks for Survival and Mobility: Social Capital Among African-American and Latin-American Low-Income Mothers." *Social Problems* 50:111–35.

Duncan, Cynthia M. 1992. *Rural Poverty in America*. Westport, CT: Auburn House.

Durkheim, Émile. 1895. *The Division of Labor in Society*. Reprint, New York: Free Press, 1964.

Eccles, Jacquelynne, and Nicole Zarrett. 2006. "The Passage to Adulthood: Challenges of Late Adolescence." *New Directions for Youth Development* 111:13–28.

Economy, Elizabeth. 2005. *The River Runs Black: The Environmental Challenge to China's Future*. Ithaca, NY: Cornell University Press.

Edmonds, Eric V. 2005. "Does Child Labor Decline With Improving Economic Status?" *Journal of Human Resources* 40:77–99.

Edwards, Richard. 1979. *Contested Terrain: The Transformation of the Workforce in the Twentieth Century*. New York: Basic Books.

Ehrenreich, Barbara. 2001. *Nickel and Dimed: On (Not) Getting By in America*. New York: Metropolitan.

Eisenbrey, Ross, and Ross Bernstein. 2003. *Eliminating the Right to Overtime Pay: Department of Labor Proposal Means Lower Pay, Longer Hours for Millions of Workers*. Washington DC: Economic Policy Institute.

Elder, Glen. 1998. "The Life Course and Human Development." In *Handbook of Child Psychology, Vol.1: Theoretical Models of Human Development,* edited by William Damon and Richard M. Lerner, 939–91. New York: Wiley.

———. 1999. *Children of the Great Depression: Social Change in Life Experience*. Boulder, CO: Westview Press.

Elger, Tony, and Chris Smith. 1994. *Global Japanization: The Transnational Transformation of the Labour Process.* London: Routledge.

Elliott, James, and Ryan Smith. 2001. "Ethnic Matching of Supervisors to Subordinate Work Groups: Findings on 'Bottom-up' Ascription and Social Closure." *Social Problems* 48:258–76.

———. 2004. "Race, Gender and Workplace Power." *American Sociological Review* 69:365–86.

Emirbayer, Mustafa, and Ann Mische. 1998. "What Is Agency?" *American Journal of Sociology* 103:962–1023.

Engels, Friederich. 1845. *The Condition of the Working-Class in England in 1844.* Reprint, New York: Allen & Unwin, 1936.

England, Paula, Michelle Budig, and Nancy Folbre. 2002. "Wages of Virtue: The Relative Pay of Care Work." *Social Problems* 49:455–73.

Epstein, David. 2005. "Fired Up at NYU." *Inside Higher Ed*, July 13.

Erikson, Kai. 1994. *A New Species of Trouble.* New York: Norton.

European Foundation for the Improvement of Living and Working Conditions. 2011. *Changes Over Time: First Findings From the Fifth European Working Conditions Survey.* Dublin: European Foundation for the Improvement of Living and Working Conditions.

Fagnani, Jeanne, and Marie-Therese Letablier. 2004. "Work and Family Life Balance: The Impact of the 35 Hour Laws in France." *Work, Employment and Society* 18:551–72.

Fantasia, Rick, and Kim Voss. 2004. *Hard Work: Remaking the American Labor Movement.* Berkeley: University of California Press.

Farnesworth-Riche, Martha. 2006. Demographic Implications for Work-Family Research. In *The Work and Family Handbook: Multi-disciplinary Perspectives, Methods, and Approaches*, edited by Marcie Pitt-Catsouphes, Ellen E. Kossek, and Stephen Sweet, 125–140. Mahwah, NJ: Lawrence Erlbaum.

Feagin, Joe. 1991. "The Continuing Significance of Race: Antiblack Discrimination in Public Places." *American Sociological Review* 56:101–16.

Feagin, Joe, and Robert Parker. 1990. *Building American Cities: The Urban Real Estate Game.* Englewood Cliffs, NJ: Prentice Hall.

Fernandez, Roberto, and Isabel Fernandez-Mateo. 2006. "Networks, Race and Hiring." *American Sociological Review* 71:42–71.

Fernandez, Roberto, and Celina Su. 2004. "Space in the Study of Labor Markets." *Annual Review of Sociology* 30:545–69.

Fine, Janice. 2006. *Worker Centers: Organizing Communities at the Edge of the Dream.* Ithaca, NY: Cornell University Press.

Fletcher, Joyce. 2001. *Disappearing Acts: Gender, Power, and Relational Practice at Work.* Cambridge: MIT Press.

Florida, Richard. 2002. *The Rise of the Creative Class: And How It's Transforming Work, Leisure, Community, and Everyday Life.* New York: Basic Books.

Fredrickson, George. 1981. *White Supremacy: A Comparative Study in American and South African History.* New York: Oxford University Press.

Freeman, Richard. 2005. "Fighting for Other Folks' Wages: The Logic and Illogic of Living Wage Campaigns." *Industrial Relations* 44:14–31.

Freud, Sigmund. 1929. *Civilization and Its Discontents*. Reprint, New York: Norton, 1961.

Freund, Peter, and George Martin. 2000. "Driving South: The Globalization of Auto Consumption and Its Social Organization of Space." *Capitalism, Nature, Socialism* 11:51–71.

Fried, Mindy. 1998. *Taking Time: Parental Leave Policy and Corporate Culture*. Philadelphia: Temple University Press.

Friedan, Betty. 1963. *The Feminine Mystique*. New York: Norton.

Friedman, Thomas. 2005. *The World Is Flat: A Brief History of the Twenty-First Century*. New York: Farrar, Straus and Giroux.

Fröbel, Folker, Jürgen Heinrichs, and Otto Kreye. 1982. *The New International Division of Labour: Structural Unemployment in Industrialised Countries and Industrialisation in Developing Countries*. New York: Cambridge University Press.

Fuchs-Epstein, Cynthia. 2007. "Great Divides: The Cultural, Cognitive, and Social Bases of the Global Subordination of Women." *American Sociological Review* 72:1–22.

Galbraith, John Kenneth. 1976. *The Affluent Society*. 3rd ed. Boston: Houghton Mifflin.

Galinsky, Ellen, and James T. Bond. 1998. *The National Study of the Changing Work Force*. New York: Families and Work Institute.

Gallant, Mary J., and Jay E. Cross. 1993. "Wayward Puritans in the Ivory Tower: Collective Aspects of Gender Discrimination in Academia." *The Sociological Quarterly* 34:237–56.

Galtry, Judith. 2000. "Extending the 'Bright Line': Feminism, Breastfeeding, and the Workplace in the United States." *Gender & Society* 14:295–317.

———. 2002. "Child Health: An Underplayed Variable in Parental Leave Policy Debates?" *Community, Work & Family* 5:257–78.

Garey, Anita Ilta. 1999. *Weaving Work and Motherhood*. Philadelphia: Temple University Press.

Garfinkel, Harold. 1967. *Studies in Ethnomethodology*. Englewood Cliffs, NJ: Prentice Hall.

Gereffi, Gary. 1994. "Capitalism, Development and Global Commodity Chains." In *Capitalism and Development*, edited by Gary Gereffi, 211–31. London: Routledge.

Gergen, Kenneth. 1991. *The Saturated Self: Dilemmas of Identity in Contemporary Life*. New York: Basic Books.

Gerson, Kathleen. 2001. "Children of the Gender Revolution: Some Theoretical Questions and Findings From the Field." In *Restructuring Work and the Life Course*, edited by Victor W. Marshall, Walter R. Heinz, Helga Krueger, and Anil Verma, 446–61. Toronto: University of Toronto Press.

———. 2009. *The Unfinished Revolution: How a New Generation Is Reshaping Family, Work, and Gender in America*. New York: Oxford.

Gerstel, Naomi, and Katherine McGonanagle. 1999. "Job Leaves and the Limits of the Family and Medical Leave Act." *Work and Occupations* 26:510–34.

Gerstel, Naomi, and Natalia Sarkisian. 2006. "Sociological Perspectives on Families and Work: The Import of Gender, Class, and Race." In *The Work and Family Handbook: Multi-disciplinary Perspectives, Methods, and Approaches*, edited by Marcie Pitt-Catsouphes, Ellen E. Kossek, and Stephen Sweet, 73–98. Mahwah, NJ: Lawrence Erlbaum.

Gilbert, Dennis. 2003. *The American Class Structure in an Age of Growing Inequality*. New York: Wadsworth.

Gitelson, I. B., and Dana McDermott. 2006. "Parents and Their Young Adult Children: Transitions to Adulthood." *Child Welfare* 85:853–66.

Giuffre, Patti, and Christine Williams. 1994. "Boundary Lines: Labeling Sexual Harassment in Restaurants." *Gender & Society* 8:378–401.

Glenn, Eveyln Nakano. 2002. *Unequal Freedom: How Race and Gender Shaped American Citizenship and Labor*. Cambridge, MA: Harvard University Press.

Glover, Katherine. 2003. "No Sweat in Minneapolis." *Dollars and Sense* 249:8–9.

Glyn, Andrew. 2005. "The Imbalance of the Global Economy." *New Left Review* 34:5–37.

Goldberg, Jeffrey 2007. "Selling Wal-Mart: Annals of Spin." *The New Yorker*, April 2.

Golden, Lonnie. 1998. *Family Friend or Foe?: Working Time, Flexibility, and the Fair Labor Standards Act*. Washington, DC: Economic Policy Institute.

———. 2001. "Flexible Work Schedules: Which Workers Get Them?" *American Behavioral Scientist* 44:1157–78.

———. 2005. "Overemployment in the US: Which Workers Face Downward Constrained Hours." In *Decent Working Time: New Trends, New Issues*, edited by Jean-Yves Boulin, Michael Lallement, Jon Messenger, and François Michon, 209–34. New York: International Labor Organization.

Goldfield, Mike. 1987. *The Decline of Organized Labor in the United States*. Chicago: University of Chicago Press.

Goldscheider, Frances, and Calvin Goldscheider. 1999. *The Changing Transition to Adulthood: Leaving and Returning Home*. Thousand Oaks, CA: Sage.

Goldscheider, Frances, Calvin Goldscheider, Patricia St. Clair, and James Hodges. 1999. "Changes in Returning Home in the United States, 1925–1985." *Social Forces* 78:695–728.

Gordon, Jennifer. 2005. *Suburban Sweatshops: The Fight for Immigrant Rights*. Cambridge, MA: Harvard University Press.

Gornick, Janet, and Marcia Meyers. 2003. *Families That Work: Policies for Reconciling Parenthood and Employment*. New York: Russell Sage Foundation.

Gorz, Andre. 1982. *Farewell to the Working Class*. Boston: South End Press.

Granovetter, Mark. 1973. "The Strength of Weak Ties." *American Journal of Sociology* 78:1360–80.

———. 1995. *Getting a Job*. Chicago: University of Chicago Press.

Gratton, Brian, and Jon Moen. 2004. "Immigration, Culture, and Child Labor in the United States, 1880–1920." *Journal of Interdisciplinary History* 34:355–91.

Green, Francis. 2006. *Demanding Work: The Paradox of Job Quality in the Affluent Economy*. Princeton, NJ: Princeton University Press.

Green, James. 2006. *Death in the Haymarket*. New York: Pantheon.

Greenhouse, Steven. 2011a. "A Watershed Moment for Public-Sector Unions." *New York Times*, February 19.

————. 2011b. "Losing Vote at a Target in New York, Union Vows to Try Again." *New York Times*, June 18.

Grunberg, Leon, Richard Anderson-Connolly, and Edward S. Greenberg. 2000. "Surviving Layoffs: The Effects on Organizational Commitment and Job Performance." *Work and Occupations* 27:7–31.

Grusky, David, and Maria Charles. 2004. *Occupational Ghettos: The Worldwide Segregation of Women and Men*. Stanford, CA: Stanford University Press.

Gusfield, Joseph. 1963. *Symbolic Crusade: Status Politics and the Urban Temperance Movement*. Chicago: University of Illinois.

Haas, Linda, and Tine Rostgaard. 2011. "Fathers' Rights to Paid Parental Leave in the Nordic Countries: Consequences for the Gendered Division of Leave." *Community, Work & Family* 14:179–97.

Hacker, Jacob. 2006. *The Great Risk Shift: The Assault on American Jobs, Families, Health Care and Retirement and How You Can Fight Back*. New York: Oxford University Press.

Hakim, Catherine. 2001. *Work-Lifestyle Choices in the Twenty-First Century*. New York: Oxford.

Hamilton, V. Lee, Clifford L. Broman, William S. Hoffman, and Deborah S. Renner. 1990. "Hard Times and Vulnerable People: Initial Effects of Plant Closing on Autoworkers' Mental Health." *Journal of Health and Social Behavior* 31:123–40.

Harden, Jeni. 2001. "Mother Russia at Work: Gender Divisions in the Medical Profession." *European Journal of Women's Studies* 8:181–99.

Hareven, Tamara. 1982. *Family Time and Industrial Time: The Relationship Between the Family and Work in a New England Industrial Community*. New York: Cambridge University Press.

Hareven, Tamara, and Randolph Langenbach. 1978. *Amoskeag*. New York: Pantheon Books.

Harrington, Brad, and Jaquelyn James. 2006. "The Standards of Excellence in Work-Life Integration: From Changing Policies to Changing Organizations." In *The Work and Family Handbook: Multi-disciplinary Perspectives, Methods, and Approaches*, edited by Marcie Pitt-Catsouphes, Ellen E. Kossek, and Stephen Sweet, 665–84. Mahwah, NJ: Lawrence Erlbaum.

Harris, William. 1982. *The Harder We Run: Black Workers Since the Civil War* New York: Oxford University Press.

Harrison, Bennett. 1997. *Lean and Mean: Why Large Corporations Will Continue to Dominate the Global Economy*. New York: Guilford Press.

Harvey, Adia. 2005. "Becoming Entrepreneurs: Intersections of Race, Class and Gender at the Black Beauty Salon." *Gender & Society* 19:789–808.

Haugen, Steven. 2003. "Characteristics of Minimum Wage Workers in 2002." *Monthly Labor Review* 126:37–40.

Hays, Sharon. 2003. *Flat Broke With Children: Women in the Age of Welfare Reform*. Cambridge, UK: Oxford University Press.

Head, Simon. 2003. *The New Ruthless Economy: Work and Power in the Digital Age*. Cambridge, UK: Oxford University Press.

———. 2004. "Inside the Leviathan." *New York Review of Books*. December 16.

Heckscher, Charles. 1988. *The New Unionism: Employee Involvement in the Changing Corporation*. New York: Basic Books.

———. 1996. *White Collar Blues: Management Loyalties in the Age of Corporate Restructuring*. New York: Basic Books.

Hegewisch, Ariane, and Janet Gornick. 2011. "The Impact of Work-Family Policies on Women's Employment: A Review of Research From OECD Countries." *Community, Work & Family* 14:119–37.

Henson, Kevin. 1996. *Just a Temp*. Philadelphia: Temple University Press.

Henson, Kevin, and Jackie Krasas Rogers. 2001. "Why Marcia You've Changed! Male Temporary Clerical Workers Doing Masculinity in a Feminized Occupation." *Gender & Society* 15:218–38.

Herrnstein, Richard, and Charles Murray. 1994. *The Bell Curve: Intelligence and Class Structure in American Life*. New York: Free Press.

Hessler, Peter. 2003. "Underwater." *The New Yorker*. July 7.

Heymann, Jody. 2000. *The Widening Gap: Why America's Working Families Are in Jeopardy and What Can Be Done About It*. New York: Perseus.

Hira, Ron. 2010a. *The H-1B and L-1 Visa Programs: Out of Control*. Washington DC: Economic Policy Institute.

———. 2010b. "U.S. Policy and the STEM Workforce System." *American Behavioral Scientist* 53:949–61.

Hirst, Paul, and J. Zeitlin. 1991. "Flexible Specialization Versus Post-Fordism." *Economy and Society* 20:1–56.

Hochschild, Arlie. 1983. *The Managed Heart: Commercialization of Human Feeling*. Berkeley: University of California Press.

———. 1997. *The Time Bind: When Work Becomes Home and Home Becomes Work*. New York: Metropolitan Books.

Hochschild, Arlie, and Anne Machung. 1989. *The Second Shift: Working Parents and the Revolution at Home*. New York: Viking Press.

Hoefer, Michael, Nancy Rytina, and Ryan Baker. 2011. *Estimates of the Unauthorized Immigrant Population Residing in the United States: January 2010*. Washington, DC: U.S. Department of Homeland Security.

Hoffman, William S., Patricia Carpenter-Alting, Duane Thomas, V. Lee Hamilton, and Clifford L. Broman. 1991. "Initial Impact of Plant Closings on Automobile Workers and Their Families." *Families in Society* 72:103–7.

Holzer, Harry 1991. "The Spatial Mismatch Hypothesis: What Has the Evidence Shown." *Urban Studies* 28:105–22.

Holzer, Harry, Julia Lane, David Rosenblum, and Fredrik Andersson. 2011. *Where Are All the Good Jobs Going? What National and Local Job Quality and Dynamics Mean for U.S. Workers*. New York: Russell Sage Foundation.

Honey, Michael 1999. *Black Workers Remember: An Oral History of Segregation, Unionism and the Freedom Struggle.* Berkeley: University of California Press.

Hostetler, Andrew, Stephen Sweet, and Phyllis Moen. 2007. "Gendered Career Paths: A Life Course Perspective on the Return to School." *Sex Roles* 56:85–103.

Hounshell, David. 1984. *From the American System to Mass Production.* Baltimore: Johns Hopkins University Press.

Hutchens, Robert M. and Emma Dentinger. 2003. "Moving Toward Retirement." In *It's About Time: Couples and Careers*, edited by Phyllis Moen, 259–74. Ithaca, NY: Cornell University Press.

Ibarra, Hermania. 1993. "Personal Networks of Women and Minorities in Management: A Conceptual Framework." *Academy of Management Review* 18:56–87.

Iversen, Lars, and Svend Sabroe. 1988. "Psychological Well-Being Among Unemployed and Employed People After a Company Closedown: A Longitudinal Study." *The Journal of Social Issues* 44:141–52.

Iwanow, H., M. G. McEachern, and A. Jeffrey. 2005. "The Influence of Ethical Trading Policies on Consumer Apparel Purchase Decisions." *International Journal of Retail and Distribution Management* 33: 371–87.

Jackall, Robert. 1989. *Moral Mazes: The World of Corporate Managers.* Cambridge, UK: Oxford University Press.

Jacobs, Jerry. 1999. "The Sex Segregation of Occupations." In *Handbook of Gender and Work*, edited by Gary N. Powell, 125–41. Thousand Oaks, CA: Sage.

Jacobs, Jerry A., and Kathleen Gerson. 2004. *The Time Divide: Work, Family, and Gender Inequality.* Cambridge, MA: Harvard University Press.

Jacoby, Sanford. 1985. *Employing Bureaucracy.* New York: Columbia University Press.

———. 1991. *Masters to Managers.* New York: Columbia University Press.

———. 1999. "Are Career Jobs Headed for Extinction?" *California Management Review* 42:123–45.

———. 2001. "Risk and the Labor Market: Societal Past as Economic Prologue." In *Sourcebook of Labor Markets: Evolving Structures and Processes*, edited by Ivar Berg and Arne Kalleberg, 31–60. New York: Kluwer.

Jahoda, Marie. 1982. *Employment and Unemployment: A Social Psychological Analysis.* New York: Cambridge University Press.

Jencks, Christopher, and Paul Peterson. 1991. *The Urban Underclass.* Washington, DC: Brookings Institution.

Johnston, David Cay. 2004. *Perfectly Legal: The Covert Campaign to Rig Our Tax System to Benefit the Super Rich—and Cheat Everybody Else.* New York: Portfolio.

Jones, Ellis, Ross Haenfler, Brett Johnson, and Brian Klocke. 2002. *The Better World Handbook: From Good Intentions to Everyday Actions.* Gabriola Island, BC: New Society.

Jones, Marc T. 2005. "The Transnational Corporation, Corporate Social Responsibility and the 'Outsourcing' Debate." *Journal of American Academy of Business, Cambridge* 6:91–7.

Kahlenberg, Richard. 2004. *America's Untapped Resource: Low-Income Students in Higher Education*. New York: Century Foundation Press.

Kahn, Joseph. 2003. "Ruse in Toyland: Chinese Workers' Hidden Woe." *New York Times*, December 7.

Kalleberg, Arne. 2007. *The Mismatched Worker*. New York: Norton.

———. 2009. "Precarious Work, Insecure Workers: Employment Relations in Transition." *American Sociological Review* 74:1–22.

Kalleberg, Arne, Barbara Reskin, and Ken Hudson. 2000. "Bad Jobs in America: Standard and Nonstandard Employment Relations and Job Quality in the United States." *American Sociological Review* 65:256–78.

Kamerman, Sheila. 2000. "Parental Leave Policies: An Essential Ingredient in Early Childhood Education and Care Policies." *Social Policy Report* 14:3–15.

Kan, Man Yee. 2007. "Work Orientation and Wives' Employment Careers: An Evaluation of Hakim's Preference Theory." *Work and Occupations* 34:430–62.

Kanter, Rosabeth Moss. 1977. *Men and Women of the Corporation*. New York: Basic Books.

Karlin, Carolyn Aman, Paula England, and Mary Richardson. 2002. "Why Do 'Women's Jobs' Have Low Pay for Their Educational Level?" *Gender Issues* 20:3–22.

Katz, Michael. 1996. *In the Shadow of the Poorhouse*. New York: Basic Books.

Kelly, Erin. 2003. "The Strange History of Employer-Sponsored Child Care: Interested Actors, Uncertainty, and the Transformation of Law in Organizational Fields." *American Journal of Sociology* 109:606–49.

———. 2006. "Work-Family Policies: The United States in International Perspective." In *The Work and Family Handbook: Multi-disciplinary Perspectives, Methods, and Approaches*, edited by Marcie Pitt-Catsouphes, Ellen E. Kossek, and Stephen Sweet, 99–124. Mahwah, NJ: Lawrence Erlbaum.

Kelly, Erin, Ellen Kossek, Leslie Hammer, Mary Durham, Jeremy Bray, Kelly Chermack, Lauren Murphy, and Dan Kaskubar. 2008. "Getting There From Here: Research on the Effects of Work-Family Initiatives on Work-Family Conflict and Business Outcomes." In *The Academy of Management Annals (Volume 2)*, edited by James Walsh and Arthur Brief, 305–49. New York: Routledge.

Kelly, Erin, Phyllis Moen, and Eric Tranby. 2011. "Changing Workplaces to Reduce Work-Family Conflict: Schedule Control in a White-Collar Organization." *American Sociological Review* 76:265–90.

Kelvin, Peter, and Joanna E. Jarrett. 1985. *Unemployment: Its Social Psychological Effects*. Cambridge: Cambridge University Press.

Kenessey, Zoltan. 1987. "Primary, Secondary, Tertiary and Quarternary Sectors of the Economy." *The Review of Income and Wealth* 33:359–86.

Kennelly, Ivy. 1999. "That Single Mother Element: How White Employers Typify Black Women." *Gender & Society* 13:168–92.

———. 2002. "'I Would Never Be a Secretary': Reinforcing Gender in Segregated and Integrated Occupations." *Gender & Society* 16:603–24.

———. 2006. "Secretarial Work, Nurturing, and the Ethic of Service." *NWSA Journal* 18:170–92.

Kets de Vries, Manfred, and Katarina Balazs. 1997. "The Downside of Downsizing." *Human Relations* 50:11–50.

Khadria, Binod. 2001. "Shifting Paradigms of Globalization: The Twenty-First Century Transition Towards Generics in Skilled Migration From India." *International Migration*. 39(5):45–71.

Kim, Chigon, and Mark Gottdiener. 2004. "Urban Problems in Global Perspective." In *Handbook of Social Problems: A Comparative International Perspective* edited by George Ritzer, 172–92. Thousand Oaks, CA: Sage.

Kimmel, Michael. 2006. "A War Against Boys?" *Dissent* 53:65-70.

Kirschenmann, Jolene, and Kathryn Neckerman. 1991. "'We'd Love to Hire Them But . . .': The Meaning of Race for Employers." In *The Urban Underclass* edited by Christopher Jencks and Paul E. Peterson, 203–32. Washington DC: Brookings Institution.

Kochan, Thomas. 2005. *Restoring the American Dream: A Working Families' Agenda for America*. Cambridge: MIT Press.

Kocieniewski, David. 2011. "G.E. Strategies Let It Avoid Taxes Altogether." *New York Times*. March 25.

Koeber, Charles. 2002. "Corporate Restructuring, Downsizing and the Middle Class: The Process and Meaning of Worker Displacement in the 'New Economy.'" *Qualitative Sociology* 25:217–46.

Kossek, Ellen Ernst, and Alyssa Fried. 2006. "The Business Case: Managerial Perspectives on Work and Family." In *The Work and Family Handbook: Multi-disciplinary Perspectives, Methods, and Approaches*, edited by Marcie Pitt-Catsouphes, Ellen E. Kossek, and Stephen Sweet, 611–26. Mahwah, NJ: Lawrence Erlbaum.

Kossek, Ellen Ernst, and Cynthia Ozeki. 1999. "Bridging the Work-Family Policy and Productivity Gap." *Community, Work & Family* 2:7–32.

Kozol, Jonathan. 2006. *The Shame of the Nation*. New York: Three Rivers Press.

Kraft, Philip. 1984. *Programmers and Managers: The Routinization of Computer Programming*. Secaucus, NJ: Springer Verlag.

Kremer, Monique. 2006a. "Consumers in Charge of Care: The Dutch Personal Budget and Its Impact on the Market, Professionals, and the Family." *European Societies* 8:385–401.

———. 2006b. "The Politics of Ideals of Care: Danish and Flemish Child Care Policy Compared." *Social Politics* 13:261–85.

Krugman, Paul. 2005. "French Family Values." *New York Times,* July 29.

Kurlansky, Mark. 1998. *Cod: A Biography of the Fish That Changed the World*. New York: Penguin.

Lafer, Gordon. 2003. "Graduate Student Unions: Organizing in a Changed Academic Economy." *Labor Studies Journal* 28:25–43.

Lambert, Susan. 2010. "Passing the Buck: Labor Flexibility Practices That Transfer Risk Onto Hourly Workers." *Human Relations* 61:1203–27.

Lasch, Christopher. 1995. *Haven in a Heartless World: The Family Besieged*. New York: Norton.

Lash, Scott, and John Urry. 1987. *The End of Organized Capitalism*. Madison: University of Wisconsin Press.

Latack, Janina, Angelo Kinicki, and Gregory Prussia. 1995. "An Integrative Process Model of Coping With Job Loss." *Academy of Management Review* 20:311–34.

Leana, Carrie R., and Daniel C. Feldman. 1992. *Coping With Loss: How Individuals, Organizations and Communities Respond to Layoffs*. New York: Lexington Books.

———. 1995. "Finding New Jobs After a Plant Closing: Antecedents and Outcomes of the Occurrence and Quality of Reemployment." *Human Relations* 48: 1381–1401.

Lee, Ching Kwan. 1998. *Gender and the South China Miracle*. Berkeley: University of California Press.

Leidner, Robin. 1993. *Fast Food, Fast Talk: Service Work and the Routinization of Everyday Life*. Los Angeles: University of California Press.

Levy, Frank. 1998. *The New Dollars and Dreams*. New York: Russell Sage.

Lichtenstein, Nelson. 2002. *State of the Union: A Century of American Labor*. Princeton, NJ: Princeton University Press.

Liebow, Eliot. 1967. *Tally's Corner*. Boston: Little, Brown.

Lim, Nelson. 2001. "On the Backs of Blacks? Immigrants and the Fortunes of African Americans." In *Strangers at the Gates: New Immigrants in Urban America*, edited by R. Waldinger, 186–227. Berkeley: University of California Press.

Lipset, Seymour. 1996. *American Exceptionalism: A Double-Edged Sword*. New York: Norton.

Liptak, Adam. 2011. "Jutices Rule for Wal-Mart in Class-Action Bias Case." *New York Times*, June 21.

Lopez, Steven. 2003. "Overcoming Legacies of Business Unionism: Why Grassroots Organizing Activities Succeed." In *Rebuilding Labor: Organizers and Organizing in the New Labor Movement*, edited by Ruth Milkman and Kim Voss, 114–32. Ithaca, NY: ILR Press.

Lorber, Judith. 1994. *Paradoxes of Gender*. New Haven: Yale University Press.

Lott, Bernice, and Diane Maluso. 1993. "The Social Learning of Gender." In *The Psychology of Gender*, edited by Anne E. Beall and Robert J. Sternberg, 99–123. New York: Guilford Press.

Lowell, Lindsay. 2010. "A Long View of America's Immigration Policy and the Supply of Foreign-Born STEM Workers in the United States." *American Behavioral Scientist* 53:1029–44.

Lowenstein, Roger. 2006. "Do Illegal Immigrants Take Jobs?" *New York Times Magazine*. July 9.

Luken, Ralph, and Rodney Stares. 2005. "Small Business Responsibility in Developing Countries: A Threat or an Opportunity?" *Business Strategy and the Environment* 14:38–53.

Machlowitz, Marilyn M. 1980. *Workaholics: Living With Them, Working With Them*. Reading, MA: Addison-Wesley.

MacKinnon, Catherine. 1998. *Sex Equality*. New York: Foundation Press.

MacLeod, Jay. 1995. *Ain't No Makin' It*. Boulder: Westview Press.

MacMillan, Ross, ed. (2005). *The Structure of the Life Course: Standardized? Individualized? Differentiated? Advances in Life Course Research, vol. 9*. New York: Elsevier.

Maher, Kris. 2011. "Unions Fend Off Right to Work Bill in New Hampshire." *Wall Street Journal*, June 24.

Mandel, Hadas. 2011. "Rethinking the Paradox: Tradeoffs in Work-Family Policy and Patterns of Gender Inequality." *Community, Work & Family* 14:159–77.

Mandel, Hadas, and Moshe Semyonov. 2004. "Family Policies, Wage Structures, and Gender Gaps: Sources of Earnings Inequality in 20 Countries." *American Sociological Review* 70:949–67.

———. 2006. "A Welfare State Paradox: State Interventions and Women's Employment Opportunities in 22 Countries." *American Journal of Sociology* 111:1910–49.

Marglin, Steven. 1982. "What Do the Bosses Do? The Origins and Functions of Hierarchy in Capitalist Production." In *Classes, Power and Conflict: Classical and Contemporary Debates*, edited by Anthony Giddens and David Held, 285–98. Berkeley: University of California Press.

Marks, Alexandra. 2004. "United's Pension Woes: Sign of a Bigger Issue." *Christian Science Monitor*, October 4.

Marshall, Ray, and Marc Tucker. 1992. *Thinking for a Living: Education and the Wealth of Nations*. New York: Basic Books.

Martin, Susan Ehrlich. 1982. *Breaking and Entering: Policewomen on Patrol*. Los Angelas: University of California Press.

Marx, Karl. 1867. *Capital*. Reprint, London: Lawrence & Wishart, 1970.

———. 1964. "Economic and Philosophic Manuscripts." In *Karl Marx: Economic and Philosophic Manuscripts*, edited by T. B. Bottomore, pp. 61–219 New York: McGraw Hill.

Marx, Karl, and Friedrich Engels. 1972. "Manifesto of the Communist Party." In *The Marx-Engels Reader*, edited by Robert C. Tucker, 331–62. New York: Norton.

Massachusetts Institute of Technology. 1999. "A Study on the Status of Women Faculty in Science at MIT." Accessed October 13, 2011. http://web.mit.edu/fnl/women/women.html.

Masser, Barbara, and Dominic Abrams. 2004. "Reinforcing the Glass Ceiling: The Consequences of Hostile Sexism for Female Managerial Candidates." *Sex Roles* 51:609–15.

Massey, Douglas. 2006. "The Wall That Keeps Illegal Workers In." *New York Times*, October 4.

Mattingly, Doreen. 1999. "Making Maids." In *Gender, Migration & Domestic Service*, edited by Janet Momsen, 61–78. New York: Routledge.

Mattsson, Lars-Gunnar. 2003. "Reorganization of Distribution in Globalization of Markets: The Dynamic Context of Supply Chain Management." *Supply Chain Management* 8:416–26.

McCann, Deirdre. 2005. *Working Time Laws: A Global Perspective: Findings From the ILO's Conditions of Work and Employment Database.* Geneva: International Labor Office.

McElhinney, Stephen. 2005. "Exposing the Interests: Decoding the Promise of the Global Knowledge Society." *New Media & Society* 7:748–69.

McPhee, John. 2005. "Annals of Transport: Out in the Sort." *New Yorker* April 18:161–67.

Meiksins, Peter. 1998. "The Time Bind." *Monthly Review* 49:1–13.

Meiksins, Peter, and Peter Whalley. 2002. *Putting Work in Its Place: A Quiet Revolution.* Ithaca, NY: ILR Press.

———. 2004. "Labor and Leisure: Should Europe Work More, or America Less?" *International Herald Tribune*, August 11.

Meyer, Steven. 1981. *The Five Dollar Day: Labor Management and Social Control in the Ford Motor Company, 1908–1921.* Albany: SUNY Press.

Meyerson, Harold. 2011. "Editorial: The U.S.: Where Europe Comes to Slum." *Los Angeles Times*, May 15.

Milanovic, Branko. 2005. *Worlds Apart: Measuring International and Global Inequality.* Princeton, NJ: Princeton University Press.

Milberg, William. 2004. "The Changing Structure of Trade Linked to Global Production Systems: What Are the Policy Implications?" *International Labour Review* 143:45–90.

Milkman, Ruth. 1997. *Farewell to the Factory: Auto Workers in the Late Twentieth Century.* Berkeley: University of California Press.

———. 2006. *L.A. Story: Immigrant Workers and the Future of the U.S. Labor Movement.* Berkeley: University of California Press.

Milkman, Ruth, and Kim Voss. 2004. *Rebuilding Labor: Organizers and Organizing in the New Labor Movement.* Ithaca, NY: ILR Press.

Mills, C. Wright. 1951. *White Collar: The American Middle Classes.* Reprint, London: Oxford University Press, 2002.

———. 1959. *The Sociological Imagination.* New York: Oxford University Press.

———. 1978. *The Power Elite.* New York: Oxford University Press.

Mir, Ali, Biju Mathew, and Raza Mir. 2000. "The Codes of Migration: Contours of the Global Software Labor Market." *Cultural Dynamics* 12:5.

Mischel, Lawrence, Jared Bernstein, and Sylvia Allegretto. 2005. *The State of Working America 2004/2005.* Ithaca, NY: Economic Policy Institute.

Mischel, Lawrence, Jared Bernstein, and Heidi Shierholz. 2009. *The State of Working America 2008–09.* Ithaca, NY: Economic Policy Institute.

Mishra, Aneil K., and Gretchen M. Spreitzer. 1998. "Explaining How Survivors Respond to Downsizing." *Academy of Management Review* 23:567–88.

Misra, Joya, Michelle Budig, and Irene Boeckman. 2011. "Work-Family Policy and the Effects of Children on Women's Employment, Hours, and Earnings." *Community, Work & Family* 14:139–57.

Moberg, David. 2005. "Look Who's Walking." *The Nation*, August 5.

Moen, Phyllis. 2001. "Constructing a Life Course." *Marriage and Family Review* 30:97–109.

Moen, Phyllis, Erin Kelly, and Rachelle Hill. 2011. "Does Enhancing Work-Time Control and Flexibility Reduce Turnover? A Naturally Occurring Experiement." *Social Problems* 58:69–98.

Moen, Phyllis, and Patricia V. Roehling. 2005. *The Career Mystique*. Boulder, CO: Rowman & Littlefield.

Moen, Phyllis, Ronit Waismel-Manor, and Stephen Sweet. 2003. "Success." In *It's About Time: Couples and Careers*, edited by Phyllis Moen, 133–52. Ithaca, NY: Cornell University Press.

Moen, Phyllis, and Stephen Sweet. 2003. "Time Clocks: Couples' Work Hour Strategies." In *It's About Time: Couples and Careers*, edited by Phyllis Moen, 17–34. Ithaca, NY: Cornell University Press.

———. 2004. "From 'Work-Family' to 'Flexible Careers': A Life Course Reframing." *Community, Work & Family* 7:209–26.

Moen, Phyllis, Stephen Sweet, and Raymond Swisher. 2005. "Embedded Career Clocks: The Case of Retirement Planning." In *The Structure of the Life Course: Standardized? Individualized? Differentiated? Advances in Life Course Research*, vol. 9, edited by Ross MacMillan, 237–268. New York: Elsevier.

Moen, Phyllis, and Elaine Wethington. 1992. "The Concept of Family Adaptive Strategies." *Annual Review of Sociology* 18:233–51.

Mohanty, Chandra Talpade. 2003. *Feminism Without Borders: Decolonizing Theory, Practicing Solidarity*. Durham, NC: Duke University Press.

Moller, Stephanie, and Beth Rubin. 2008. "The Contours of Stratification in Service-Oriented Economies." *Social Science Research* 37:1039–60.

Mong, Sherry, and Vincent Roscigno. 2010. "African American Men and the Experience of Employment Discrimination." *Qualitative Sociology* 33:1–21.

Montgomery, David. 1979. *Workers' Control in America: Studies in the History of Work, Technology, and Labor Struggles*. Cambridge: Cambridge University Press.

Moody, Kim. 1997. *Workers in a Lean World*. New York: Verso Books.

Moore, Thomas. 1996. *The Disposable Work Force: Worker Displacement and Employment Instability in America*. New York: Aldine De Gruyter.

Morgan, Kimberly. 2005. "The 'Production' of Child Care: How Labor Markets Shape Social Policy and Vice Versa." *Social Politics* 12:243–63.

Morgan, Kimberly, and Kathrin Zippel. 2003. "Paid to Care: The Origins and Effects of Care Leave Policies in Western Europe." *Social Politics* 10:49–85.

Moss, Philip, and Chris Tilly. 2001. *Stories Employers Tell: Race, Skill and Hiring in America*. New York: Russell Sage Foundation.

Mouw, Ted. 2002. "Are Black Workers Missing the Connection? The Effect of Spatial Distance and Employee Referrals on Interfirm Racial Segregation." *Demography* 39:507–28.

Mouw, Ted, and Arne Kalleberg. 2010. "Occupations and the Structure of Wage Inequality in the United States, 1980s to 2000s." *American Sociological Review* 75:402–31.

Murray, Charles. 1995. *Losing Ground: American Social Policy, 1950–1980*. New York: Basic Books.

Myrdal, Gunnar. 1954. *An American Dilemma: The Negro Problem and Modern Democracy.* Reprint, New York: Transaction, 1995.

Naumann, Stefanie, Nathan Bennett, Robert Beis, and Christopher Martin. 1998. "Laid Off, but Still Loyal." *International Journal of Conflict Management* 9:356–68.

Neal, Margaret, and Leslie Hammer. 2006. *Working Couples Caring for Children and Aging Parents: Effects on Work and Well-Being.* Mahwah, NJ: Lawrence Erlbaum.

Nee, Victor, Jimy Sanders, and Scott Sernau. 1994. "Job Transitions in an Immigrant Metropolis: Ethnic Boundaries and the Mixed Economy" *American Sociological Review* 59:849–72.

Nelson, Daniel. 1980. *Frederick W. Taylor and the Rise of Scientific Management.* Madison: University of Wisconsin Press.

Nelson, Valerie, Adrienne Martin, and Joachim Ewert. 2005. "What Difference Can They Make? Assessing the Social Impact of Corporate Codes of Practice." *Development in Practice* 15:539–45.

Nestle, Marion. 2003. *Food Politics.* Los Angeles: University of California Press.

Neumark, David. 2000. *On the Job: Is Long-Term Employment a Thing of the Past?* New York: Russell Sage Foundation.

Newburger, Eric, and Thomas Gryn. 2009. "The Foreign-Born Labor Force in the United States: 2007" (Vol. ACS-10). Washington DC: U.S. Census Bureau.

Newman, Katherine. 1999. *No Shame in My Game: The Working Poor in the Inner City.* New York: Alfred A. Knopf and Russell Sage Foundation.

———. 2006. *Chutes and Ladders.* New York: Russell Sage Foundation.

Ngai, Pun. 2005. *Made in China: Factory Workers in a Global Workplace.* Durham, NC: Duke University Press.

Nippert-Eng, Christine. 1996. *Home and Work.* Chicago: University of Chicago.

Noble, David. 1979. *America by Design: Science, Technology, and the Rise of Corporate Capitalism.* New York: Alfred A. Knopf.

———. 1984. *Forces of Production.* New York: Alfred A. Knopf.

Nowicki, Carol. 2003. "Family and Medical Leave Act." In *Sloan Work and Family Encyclopedia.* Accessed October 13, 2011. http://wfnetwork.bc.edu/encyclopedia_entry.php?id=234&area=All

O'Brien, Helena, and Sarita Gupta. 2005. "Local Power Can Change Wal-Mart: The ACORN and Jobs With Justice Organizing Strategy." *Social Policy* 36:16–20.

O'Leary, Christopher, and Stephen Wandner. 1997. "Summing Up: Achievements, Problems and Prospects." In *Unemployment Insurance in the United States: Analysis of Policy Issues,* edited by Christopher O'Leary and Stephan Wandner, 669–722. New York: W.E. Upjohn Institute.

O'Reilly, Brian 1994. "The New Deal: What Companies and Employees Owe One Another." *Fortune* 129:44–50.

Ochs, Elinor, Anthony Graesch, Angela Mittmann, Thomas Bradbury, and Rena Repetti. 2006. "Video Ethnography and Ethnoarcheaological Tracking." In *The Work and Family Handbook: Multi-disciplinary Perspectives, Methods, and Approaches,* edited by Marcie Pitt-Catsouphes, Ellen E. Kossek, and Stephen Sweet, 387–410. Mahwah NJ: Lawrence Erlbaum.

Omi, Michael, and Howard Winant. 1986. *Racial Formations in the United States: From the 1960s to the 1980s.* London: Routledge.

Osnowitz, Debra. 2010. *Freelancing Expertise: Contract Professionals in the New Economy.* Ithaca, NY: ILR Press.

Osterman, Paul. 2001. *Securing Prosperity.* Princeton, NJ: Princeton University Press.

Osterud, Nancy G. 1987. "She Helped Me Hay It as Good as a Man: Relations Among Women and Men in an Agricultural Community." In *"To Toil the Livelong Day": America's Women at Work, 1780–1980,* edited by Carol Groneman and Mary Beth Norton, 87–97. Ithaca, NY: Cornell University Press.

Padavic, Irene, and Barbara Reskin. 2002. *Women and Men at Work.* Thousand Oaks, CA: Pine Forge Press.

Pager, Devah. 2003. "The Mark of a Criminal Record." *American Journal of Sociology* 108:937–75.

Pager, Devah, Bruce Western, and Bart Bonikowski. 2009. "Discrimination in a Low-Wage Labor Market: A Field Experiment." *American Sociological Review* 74:777–99.

Parker, Mike. 1985. *Inside the Circle: A Union Guide to QWL.* Boston: South End Press.

Parker, Mike, and Jane Slaughter. 1988. *Unions and the Team Concept.* Boston: South End Press.

Parks, Virginia. 2011. "Revisiting Shibboleths of Race and Urban Economy: Black Employment in Manufacturing and the Public Sector Compared 1950–2000." *International Journal of Urban and Regional Research* 35:110–29.

Parthasarathy, Balaji. 2004. "India's Silicon Valley or Silicon Valley's India? Socially Embedding the Computer Software Industry in Bangalore." *International Journal of Urban and Regional Research* 28:664–85.

Patni, Ambika. 1999. "Silicon Valley of the East." *Harvard International Review* 21:8.

Paules, Greta Foff. 1991. *Dishing It Out: Power and Resistance Among Waitresses in a New Jersey Restaurant.* Philadelphia: Temple University Press.

Pavalko, Eliza, and Kathryn Henderson. 2006. "Combining Care Work and Paid Work: Do Workplace Policies Make a Difference." *Research on Aging* 28:359–74.

Perpitone, Julianne 2010. "More Wives Outearning Their Husbands." *CNN Money.com,* January 19.

Perrucci, Carolyn C. 1994. "Economic Strain, Family Structure and Problems With Children Among Displaced Workers." *Journal of Sociology and Social Welfare* 21:79–91.

Perrucci, Carolyn C., Robert Perrucci, and Dena B. Targ. 1997. "Gender Differences in the Economic, Psychological and Social Effects of Plant Closings in an Expanding Economy." *The Social Science Journal* 34:217–33.

Perrucci, Robert, and Carl Wysong. 2002. *The New Class Society: Goodbye American Dream?* New York: Rowman & Littlefield.

Pettit, Becky, and Jennifer Hook. 2005. "The Structure of Women's Employment in Comparative Perspective." *Social Forces* 84:779–801.

Pfau-Effinger, Birgit. 2004. *Development of Culture, Welfare States and Women's Employment in Europe*. Burlington, VT: Ashgate.

Pietrykowski, Bruce. 1999. "Beyond the Fordist/Post-Fordist Dichotomy: Working Through the Second Industrial Divide." *Review of Social Economy* 57: 177–98.

Piore, Michael. 1977. "The Dual Labor Market and Its Implications." In *Problems in Political Economy*, edited by David M. Gordon, 91–5. Lexington, MA: D.C. Heath.

Piore, Michael, and Charles Sabel. 1984. *The Second Industrial Divide: Possibilities for Prosperity*. New York: Basic Books.

Pitt-Catsouphes, Marcie, Ellen Ernst Kossek, and Stephen Sweet, eds. 2006. *The Work and Family Handbook: Multi-disciplinary Perspectives, Methods, and Approaches*. Mahwah, NJ: Lawrence Erlbaum.

Pixley, Joy. 2008. "Career-Prioritizing in Dual Earner Couples." In *Work and Family Encyclopedia*. Accessed October 11, 2011. http://wfnetwork.bc.edu/encyclopedia_entry.php?id=6473&area=All.

Polanyi, Karl. 1944. *The Great Transformation*. New York: Farrar & Rinehart.

Polivka, Anne E., and Thomas Nardone. 1989. "On the Definition of 'Contingent Work'". *Monthly Labor Review* 112:9–16.

Pollert, Anna. 1988. "Dismantling Flexibility." *Capital and Class* 34:42–75.

Porter, Eduardo. 2006. "Cost of Illegal Immigration May Be Less Than Meets the Eye." *New York Times*, April 16.

Portes, Alejandro, and John Walton. 1981. *Labor, Class, and the International System*. New York: Academic Press.

Portes, Alejandro, and Ruben Rumbaut. 2001. *Legacies: The Story of the Immigrant Second Generation*. Berkeley: University of California Press.

Portes, Alejandro, and Min Zhou. 1993. "The New Second Generation: Segmented Assimilation and Its Variants." *Annals of the American Academy of Political and Social Sciences* 530:74–96.

Poster, Winifred, and Srirupa Prasad. 2005. "Work-Family Relations in Transnational Perspective: A View From High-Tech Firms in India and the United States." *Social Problems* 52:122–46.

Presser, Harriet B. 2000. "Nonstandard Work Schedules and Marital Instability." *Journal of Marriage and the Family* 62:93–110.

———. 2003. *Working in a 24/7 Economy: Challenges for American Families*. New York: Russel Sage Foundation.

———. 2008. "Nonstandard Work Schedules Over the Life Course: A First Look." Paper presented to the American Sociological Association annual meeting, Boston, July 31.

Preston, Anne. 2004. *Leaving Science: Occupational Exit From Scientific Careers*. New York: Russell Sage Foundation.

Preston, Julia. 2011. "A Crackdown on Employing Illegal Workers." *New York Times*, May 29.

"Public Sector Unions: The Battle Ahead." 2011. *The Economist,* January 6. Accessed October 8, 2011. http://www.economist.com/node/17851305.

Putnam, Robert. 2000. *Bowling Alone: The Collapse and Revival of American Community.* New York: Simon & Schuster.

Raider, Holly, and Ronald Burt. 1996. "Boundaryless Careers and Social Capital." In *The Boundaryless Career: A New Employment Principle for a New Organizational Era,* edited by Michael B. Arthur and Denise M. Rousseau, 187–200. New York: Oxford.

Reich, Robert. 2002. *The Future of Success: Working and Living in the New Economy.* New York: Vintage.

Reid, Peter. 1989. "How Harley Beat Back the Japanese." *Fortune* 120:155–59.

Reskin, Barbara 1998. *The Realities of Affirmative Action in Employment.* Washington, DC: The American Sociological Association.

Reskin, Barbara, Debra McBrier, and Julie Kmec. 1999. "The Determinants and Consequences of Workplace Sex and Race Composition." *Annual Review of Sociology* 25:335–61.

Reskin, Barbara, and Patricia Roos. 1991. *Job Queues, Gender Queues.* Philadelphia: Temple University Press.

Richardson, Pete. 2006. "The Anthropology of the Workplace and the Family." In *The Work and Family Handbook: Multi-disciplinary Perspectives, Methods and Approaches,* edited by Marcie Pitt-Catsouphes, Ellen E. Kossek, and Stephen Sweet, 165–88. Mahwah, NJ: Lawrence Erlbaum.

Richman, Amy, Janet Civian, Laurie Shannon, Jeffrey Hill, and Robert Brennan. 2008. "The Relationship of Pereived Flexibility, Supportive Work-Life Policies, and Use of Formal Flexible Arrangements and Occasional Flexibility to Employee Engagement and Expected Retention." *Community, Work & Family* 11:183–97.

Riesman, David. 2001. *The Lonely Crowd: A Study of the Changing American Character.* New Haven, CT: Yale University Press.

Riesman, David, Nathan Glazer, and Denney Reuel. 2001. *The Lonely Crowd: A Study of the Changing American Character.* New Haven, CT: Yale University Press.

Rifkin, Jeremy. 2004. *The End of Work.* New York: Jeremy P. Tarcher/Putnam Books.

Riley, Matilda White, and John W. Riley, Jr. 2000. "Age Integration: Conceptual and Historical Background." *The Gerontologist* 40:266–70.

Rimer, Sara. 2007. "For Girls, It's Be Yourself, and Be Perfect, Too." *New York Times,* April 1.

Rinehart, James, Christopher Huxley, and David Robertson. 1997. *Just Another Car Factory? Lean Production and Its Discontents.* Ithaca, NY: ILR Press.

Ritzer, George. 2011. *The McDonaldization of Society.* Thousand Oaks, CA: Pine Forge Press.

Roberts, Bryan. 2005. "Globalization and Latin American Cities." *International Journal of Urban and Regional Research* 29:110–23.

Robinson, John, and Geoffrey Godbey. 1997. *Time for Life. The Surprising Ways Americans Use Their Free Time.* University Park: Pennsylvania State University Press.

Roy, D. 1955. "Efficiency and 'The Fix'". *American Journal of Sociology* 60: 255–66.

Royster, Deirdre. 2003. *Race and the Invisible Hand: How White Networks Exclude Black Men From Blue Collar Jobs*. Berkeley: University of California Press.

Rubin, Beth A., and Charles Brody. 2005. "Contradictions of Commitment in the New Economy: Insecurity, Time, and Technology." *Social Science Research* 34:843–61.

Rudnyckyj, Daromir. 2004. "Technologies of Servitude: Governmentality and Indonesian Transnational Labor Migration." *Anthropological Quarterly* 77:407–34.

Rybczynski, Witold. 1991. *Waiting for the Weekend*. New York: Viking Press.

Sabel, Charles, and J. Zeitlin. 1985. "Historical Alternatives to Mass Production." *Past and Present* 108:133–75.

Sahlins, Marshall. 1972. *Stone Age Economics*. Chicago: Aldine-Atherton.

Saiz, Albert. 2003. "The Impact of Immigration on American Cities: An Introduction to the Issues." *Business Review* Q4:4–23.

Sales, Esther. 1995. "Surviving Unemployment: Economic Resources and Job Loss Duration in Blue-Collar Households." *Social Work* 40:483–94.

Sampson, Robert, Jeffrey D. Morenoff, and Thomas Gannon-Rowley. 2002. "Assessing 'Neighborhood Effects': Social Processes and New Directions in Research." *Annual Review of Sociology* 51:443–78.

Sampson, Robert, S. W. Raudenbush, and Felton Earls. 1997. "Neighborhoods and Violent Crime: Multilevel Study of Collective Efficacy." *Science* 277:918–24.

Santorum, Rick. 2006. *It Takes a Family: Conservatism and the Common Good*. New York: Intercollegiate Studies Institute.

Sarkesian, Natalia, and Naomi Gerstel. 2004. "Explaining the Gender Gap in Help to Parents: The Importance of Employment." *Journal of Marriage and Family* 66:431–51.

Sassen, Saskia. 1995. "Immigration and Local Labor Markets." In *The Economic Sociology of Immigration: Essays on Networks, Ethnicity, and Entrepreneurship*, edited by Alejandro Portes, 87–127. New York: Russell Sage Foundation.

Sassen, Saskia, and R. C. Smith. 1992. "Post-industrial Growth and Economic Reorganization: The Impact on Immigrant Employment." In *United States-Mexico Relations: Labour Market Interdependence*, edited by Jorge A. Bustamante, Clark W. Reynolds, and Raul A. Honojosa Ojeda, 372–96. Stanford: Stanford University Press.

Saunders, Doug. 2010. *Arrival Cities: How the Largest Migration in History Is Reshaping Our World*. New York: Pantheon Books.

Saxenian, Anna Lee. 1996. "Beyond Boundaries: Open Labor Markets and Learning in Silicon Valley." In *The Boundaryless Career: A New Employment Principle for a New Organization Age*, edited by Michael B. Arthur and Denise M. Rousseau, 23–39. New York: Oxford.

Schiebinger, Londa, Andrea Davies Henderson, and Shannon Gilmartin. 2008. *Dual-Career Academic Couples: What Universities Need to Know*. Stanford, CA: Stanford University.

Schlosser, Eric. 2005. *Fast Food Nation*. New York: Harper Perennial.

Schmidt, Stefanie. 2000. "Job Security Beliefs in the General Social Survey: Evidence on Long-Run Trends and Comparability With Other Surveys." In *On the Job: Is Long-Term Employment a Thing of the Past*, edited by David Neumark, 300–32. New York: Russell Sage Foundation.

Schoeni, Robert, and Karen Ross. 2005. "Material Assistance Received From Families During the Transition to Adulthood." In *On the Frontier of Adulthood: Theory, Research, and Public Policy*, edited by Richard A. Settersten, Jr., Frank F. Furstenberg, Jr., and Rubén G. Rumbaut, 396–416. Chicago: University of Chicago Press.

Schor, Juliet. 1991. *The Overworked American: The Unexpected Decline of Leisure*. New York: Basic Books.

————. 1998. *The Overspent American: Upscaling, Downshifting, and the New Consumer*. New York: Basic Books.

Schrage, Elliot. 2004. "Supply and the Brand." *Harvard Business Review* 82:20–1.

Schrank, Andrew. 2004. "Ready-to-Wear Development? Foreign Investment, Technology Transfer, and Learning by Watching in the Apparel Trade." *Social Forces* 83:123–56.

Schultz, Derek. 2006. "Myths and Realities About High Tech Work." In *Surviving in the New Economy*, edited by John Amman, Tris Carpenter, and Gina Neff, 15–32. Boulder, CO: Paradigm.

Schuman, Howard, Charlotte Steeh, Lawrence Bobo, and Maria Krysan. 1998. *Racial Attitudes in America: Trends and Interpretations* (Rev. ed.). Cambridge, MA: Harvard University Press.

Schumpeter, Joseph. 1989. *Essays: On Entrepreneurs, Innovations, Business Cycles, and the Evolution of Capitalism*. New Brunswick, NJ: Transaction.

Seidman, Gay. 2007. *Beyond the Boycott: Labor Rights, Human Rights and Transnational Activism*. New York: Russell Sage Foundation.

Sennett, Richard. 1998. *The Corrosion of Character: The Personal Consequences of Work in the New Capitalism*. New York: Norton.

Shaiken, Harley. 1984. *Work Transformed: Automation and Labor in the Computer Age*. New York: Lexington Books.

Shamir, Boas. 1986a. "Protestant Work Ethic, Work Involvement and the Psychological Impact of Unemployment." *Journal of Occupational Behaviour* 7:25–38.

————. 1986b. "Unemployment and Household Division of Labor." *Journal of Marriage and the Family* 48:195–206.

Shapiro, Thomas. 2004. *The Hidden Cost of Being African American: How Wealth Perpetuates Inequality*. New York: Oxford University Press.

Shephard, Roy, and Jean Bonneau. 2002. "Assuring Gender Equity in Recruitment Standards for Police Officers." *Canadian Journal of Applied Physiology* 27:263–95.

Shih, Johanna. 2004. "Project Time in Silicon Valley." *Qualitative Sociology* 27:223–45.

Shulman, Beth. 2005. *The Betrayal of Work: How Low Wage Jobs Fail 30 Million Americans and Their Families*. New York: New Press.

Skocpol, Theda. 1992. *Protecting Soldiers and Mothers: The Political Origins of Social Policy in the United States*. Boston: Harvard University Press.

Smith, Brian T., and Beth A. Rubin. 1997. "From Displacement to Reemployment: Job Acquisition in the Flexible Economy." *Social Science Research* 26:292–308.

Smith, Chris, and Peter Meiksins. 1995. "System Society and Dominance Effects in Cross-National Organisational Analysis." *Work, Employment and Society* 9:241–61.

Smith, Craig. 2006. "Letter From Paris: 4 Simple Rules for Firing an Employee in France." *New York Times,* March 28.

Smith, Ryan. 1997. "Race, Income and Authority at Work: A Cross-Temporal Analysis of Black and White Men, 1972–1994." *Social Problems* 44:19–37.

Smith, Vicki. 1990. *Managing in the Corporate Interest: Control and Resistance in an American Bank*. Berkeley: University of California Press.

———. 2002. *Crossing the Great Divide: Worker Risk and Opportunity in the New Economy*. Ithaca, NY: Cornell University Press.

Snipp, Matthew, and Gene Summers. 1992. "American Indians and Economic Poverty." In *Rural Poverty in America*, edited by Cynthia M. Duncan, 155–76. New York: Auburn House.

Sobel, Mechal. 1987. *The World They Made Together: Black and White Values in Eighteenth-Century Virginia*. Princeton, NJ: Princeton University Press.

Stack, Carol B. 1997. *All Our Kin: Strategies for Survival in a Black Community*. New York: Basic Books.

Stainback, Kevin, Corre Robinson, and Donald Tomaskovic-Devey. 2005. "Race and Workplace Integration: A Politically Mediated Process?" *American Behavioral Scientist* 48:1200–28.

Steele, Shelby. 1990. *The Content of Our Character: A New Vision of Race in America*. New York: HarperCollins.

Stone, Pamela. 1997. *Opting Out? Why Women Really Quit Careers and Head Home*. Los Angeles: University of California Press.

Sullivan, Mercer. 1989. *Getting Paid: Youth, Crime and Work in the Inner City*. Ithaca, NY: Cornell University Press.

Sun, Huey-Lian, and Alex P. Tang. 1998. "The Intra-industry Information Effect of Downsizing Announcements." *American Business Review* 16(2):68–76.

Swanberg, Jenner, Marcie Pitt-Catsouphes, and Krista Drescher-Burke. 2005. "A Question of Justice: Disparities in Employee's Access to Flexible Schedule Arrangements." *Journal of Family Issues* 26:866–95.

Sweet, Stephen. 2007. "The Older Worker, Job Insecurity and the New Economy." *Generations* 31:45–9.

———. 2009. "When Is a Person Too Young or Too Old to Work?: Cultural Variations in Europe" (Global Issue Brief No. 2). Boston: Sloan Center on Aging and Work.

Sweet, Stephen, and Phyllis Moen. 2004. "Intimate Academics: Coworking Couples in Two Universities." *Innovative Higher Education* 28:252–74.

———. 2006. "Advancing a Career Focus on Work and Family: Insights From the Life Course Perspective." In *The Work and Family Handbook:*

Multi-disciplinary Perspectives, Methods, and Approaches, edited by Marcie Pitt-Catsouphes, Ellen E. Kossek, and Stephen Sweet, 189–208. Mahwah, NJ: Lawrence Erlbaum.

———. 2007. "Integrating Educational Careers in Work and Family: Women's Return to School and Family Life Quality." *Community, Work & Family* 10:233–52.

Sweet, Stephen, Phyllis Moen, and Peter Meiksins. 2007. "Dual Earners in Double Jeopardy: Preparing for Job Loss in the New Risk Economy." In *Workplace Temporalities, Volume 17: Research in the Sociology of Work*, edited by Beth Rubin, 437–61. New York: Elsevier.

Sweet, Stephen, Raymond Swisher, and Phyllis Moen. 2005. "Selecting and Assessing the Family Friendly Community: Adaptive Strategies of Middle Class Dual-Earner Couples." *Family Relations* 54:596–606.

Swidler, Ann. 1986. "Culture in Action: Symbols and Strategies." *American Sociological Review* 51:273–86.

Swisher, Raymond, Stephen A. Sweet, and Phyllis Moen. 2004. "The Family-Friendly Community and Its Life Course Fit for Dual-Earner Couples." *Journal of Marriage and Family* 66:281–92.

Takaki, Ronald. 2008. *A Different Mirror: A History of Multicultual America*. Boston: Back Bay Books.

Tarnoff, Curt, and Larry Nowels. 2004. *Foreign Aid: An Introductory Overview of U.S. Programs and Policy*. Washington DC: Congressional Research Service.

Tax Policy Center. 2011. "Historical Marginal Effective Tax Rates on Capital Income." Accessed October 9, 2011. http://www.taxpolicycenter.org/taxfacts/displayafact.cfm?Docid=323.

Taylor, Frederick Winslow. 1911. *The Principles of Scientific Management*. Reprint, New York: Harper, 1964.

Taylor, Phil, and Peter Bain. 2005a. "Call Centre Offshoring to India: The Revenge of History?" *Labour & Industry* 14:15–38.

———. 2005b. "India Calling to the Far Away Towns: The Call Centre Labour Process and Globalization." *Work, Employment & Society* 19:261–282.

Terrell, Nicholas. 2007. "High-Tech Jobs for a High-Tech Economy." *Occupational Outlook Quarterly* Spring:26–33.

Thernstrom, Stephan. 1980. *Poverty and Progress*. Cambridge, MA: Harvard University Press.

Thomas, William I., and Florian Znaniecki. 1958. *The Polish Peasant in Europe and America*. New York: Dover.

Thomas, William and Dorothy Swaine Thomas. 1928. *The Child in America*. Chicago: University of Chicago Press.

Thompson, E. P. 1963. *The Making of the English Working Class*. New York: Pantheon Books.

———. 1967. "Time, Work-Discipline, and Industrial Capitalism." *Past and Present* 38:56–97.

Thompson, James. 1967. *Organizations in Action*. New York: McGraw-Hill.

Thurow, Lester. 1992. *Head to Head: The Coming Economic Battle Among Japan, Europe and America.* New York: William Morrow.

Thurow, Lester C. 1999. "Jobless Figures Deceptive." *The Boston Globe.* April 20, C4.

Tiano, Susan. 1994. *Patriarchy on the Line: Labor, Gender and Ideology in the Mexican Maquila Industry.* Philadelphia: Temple University Press.

Tilly, Chris. 1995. *Half a Job: Good and Bad Part-Time Jobs in a Changing Labor Market.* Philadelphia: Temple University Press.

Tocqueville, Alexis de. 1836. *Democracy in America.* Reprint, New York: Doubleday, 1969.

Tracy, James F. 1999. "Whistle While You Work: The Disney Company and the Global Division of Labor." *Journal of Communication Inquiry* 23:374.

Trattner, Walter. 1999. *From Poor Law to Welfare State.* New York: Free Press.

Trefalt, Spela. 2010. "Interpersonal Aspects of Justice in Workplace Flexibility Enactment." In *Work and Family Encyclopedia.* Accessed October 11, 2011. http://wfnetwork.bc.edu/encyclopedia_entry.php?id=16766&area=All.

Tseming, Yang. 2004. *The Effectiveness of the NAFTA Environmental Side Agreement's Citizen Submission Process: A Case Study of the Metales y Derivados Matter.* Rochester, NY: Social Science Research Network.

Tucker, James. 1993. "Everyday Forms of Employee Resistance." *Sociological Forum* 8:25–46.

Tucker, Lee. 2000. *Fingers to the Bone: United States Failure to Protect Child Farmworkers.* Washington, DC: Human Rights Watch.

Tulin, Roger. 1984. *A Machinist's Semi-automated Life.* San Pedro, CA: Singlejack Books.

Uchitelle, Louis. 2006. *The Disposable American: Layoffs and Their Consequences.* New York: Alfred A. Knopf.

———. 2011. "From Two Breadwinners to One." *The Nation,* May 23.

Uggen, Christopher, and Amy Blackstone. 2004. "Sexual Harassment as a Gendered Expression of Power." *American Sociological Review* 69:64–92.

Ulrich, Laurel. 1982. *Good Wives: Image and Reality in the Lives of Women in Northern New England 1650–1750.* New York: Oxford University Press.

United Nations. 1989. "Convention on the Rights of the Child." Accessed October 19, 2011. http://www2.ohchr.org/english/law/crc.htm

United Nations 2005. *Population Challenges and Development Goals.* New York: United Nations.

United Nations Statistics Division. 2011. "Demographic and Social Indicators: Statistics and Indicators on Women and Men." Accessed October 13, 2011. http://unstats.un.org/unsd/demographic/products/indwm/.

United States Bureau of Labor Statistics. 2008. *Issues in Labor Statistics: Involuntary Part-Time Work on the Rise.* Washington DC: U.S. Department of Labor, Bureau of Labor Statistics.

———. 2009. *Occupational Outlook Handbook.* Washington DC: U.S. Department of Labor, Bureau of Labor Statistics.

———. 2010. *Women in the Labor Force: A Data Book.* Washington DC: U.S. Department of Labor, Bureau of Labor Statistics.

————. 2011. "Economic News Release." Washington DC: U.S. Department of Labor.

United States Census Bureau. 2010. *Statistical Abstract of the United States.* Washington, DC: U.S. Department of Commerce, Bureau of Census.

United States Census Bureau. 2011. *Income, Poverty, and Health Insurance Coverage in the United States: 2009.* Washington, DC: U.S. Department of Commerce, Bureau of Census.

United States Department of Homeland Security. 2010. *Yearbook of Immigration Statistics 2009.* Washington, DC: U.S Department of Homeland Security, Office of Immigration Statistics.

United States Department of Labor. 2007. *Unemployment Compensation: A Federal-State Partnership.* Washington DC: U.S. Department of Labor.

Valian, Virginia. 1998. *Why So Slow? The Advancement of Women.* Cambridge: MIT Press.

Vallas, Steven. 2003a. "The Adventures of Managerial Hegemony: Teamwork, Ideology, and Worker Resistance." *Social Problems* 50:204–25.

————. 2003b. "Rediscovering the Color Line Within Work Organizations: The 'Knitting or Racial Groups' Revisited." *Work and Occupations* 30:379–400.

Vallas, Steven, and John Beck. 1996. "The Transformation of Work Revisited: The Limits of Flexibility in American Manufacturing." *Social Problems* 43:339–61.

Valletta, Robert. 2000. "Declining Job Security." In *On the Job: Is Long-Term Employment a Thing of the Past?*, edited by David Neumark, 227–56. New York: Russell Sage Foundation.

Varma, Roli. 2006. *Harbingers of Global Change: India's Techno-immigrants in the United States.* Lanham, MA: Lexington Books.

Vaughan, Diane. 2006. "Air Traffic Control Today: Politics, Labor History, and Cultural Reproduction." In *Critical Solidarity: Newsletter of the Labor and Labor Movements Section of the American Sociological Association* 6(2).

Veblen, Thorstein. 1899. *The Theory of the Leisure Class.* Reprint, New York: Penguin Classics, 1994.

Vosler, Nancy R., and Deborah Page-Adams. 1996. "Predictors of Depression Among Workers at the Time of a Plant Closing." *Journal of Sociology and Social Welfare* 23:25–42.

Voydanoff, Patricia. 2007. *Work, Family, and Community: Exploring Interconnections.* Mahwah, NJ: Lawrence Erlbaum.

Wagar, Terry H. 2001. "Consequences of Work Force Reduction." *Journal of Labor Research* 22:851–62.

Waldinger, Roger. 1996. *Still the Promised City: African-Americans and the New Immigrants in Postindustrial New York.* Cambridge, MA: Harvard University Press.

————. 2001. "Up From Poverty? 'Race,' Immigration and the Fate of Low-Skilled Workers." In *Strangers at the Gates: New Immigrants in Urban America*, edited by Roger Waldinger, 80–116. Berkeley: University of California Press.

Waldinger, Roger, and Claudia Der-Martirosian. 2000. "Immigrant Workers and American Labor: Challenge . . . or Disaster?" In *Organizing Immigrants: The*

Challenge for Unions in Contemporary California, edited by Ruth Milkman, 49–80. Ithaca, NY: Cornell University Press.

———. 2001. "The Immigrant Niche: Pervasive, Persistent, Diverse." In *Strangers at the Gates: New Immigrants in Urban America*, edited by R. Waldinger, 228–71. Berkeley: University of California Press.

Waldinger, Roger, and Michael Lichter. 2003. *How the Other Half Works: Immigration and the Social Organization of Labor*. Berkeley: University of California Press.

Waldinger, Roger, Nelson Lim, and David Cort. 2007. "Bad Jobs, Good Jobs, No Jobs? The Employment Experience of the Mexican-American Second Generation." *Journal of Ethnic and Migration Studies* 33:1–35.

Wallerstein, Immanuel. 1979. *The Capitalist World Economy*. Cambridge: Cambridge University Press.

———. 1983. *Historical Capitalism*. London: Verso Books.

Wanberg, Connie R., L. W. Bunce, and Mark B. Gavin. 1999. "Perceived Fairness of Layoffs Among Individuals Who Have Been Laid Off." *Personnel Psychology* 52(1):59–84.

Wandner, Stephen, and Andrew Stettner. 2000. "Why Are Many Jobless Workers Not Applying for Benefits." *Monthly Labor Review* 123:21–32.

Warren, Elizabeth, and Amelia Warren Tyagi. 2003. *The Two-Income Trap: Why Middle-Class Mothers and Fathers Are Going Broke*. New York: Basic Books.

Weber, Max. 1905. *The Protestant Ethic and the Spirit of Capitalism*. Reprint, Los Angeles: Roxbury, 1998.

Weller, Christian, and Edward Wolff. 2005. *Retirement Income: The Crucial Role of Social Security*. Washington, DC: Economic Policy Institute.

Wells, Miriam. 2000. "Immigration and Unionization in the San Francisco Hotel Industry." In *Organizing Immigrants: The Challenge for Unions in Contemporary California*, edited by Ruth Milkman, 109–29. Ithaca, NY: Cornell University Press.

Wenneras, Christine, and Agnes Wold. 1997. "Nepotism and Sexism in Peer-Review." *Nature* 387:341–43.

Westman, Mina, Dalia Etzion, and Esti Danon. 2001. "Job Insecurity and Crossover of Burnout in Married Couples." *Journal of Organizational Behavior* 22: 467–81.

Whaples, Robert. 2010. "Hours of Work in U.S. History." In *EH.net Encyclopedia*. Accessed October 11, 2011. http://eh.net/encyclopedia/article/whaples.work. hours.us.

Wharton, Amy. 1999. "The Psychosocial Consequences of Emotional Labor." *Annals of the American Academy of Political and Social Sciences* 56:158–76.

Wharton, Carol S. 1994. "Finding Time for the 'Second Shift': The Impact of Flexible Work Schedules on Women's Double Days." *Gender & Society* 8:189–205.

———. 2002. *Framing a Domain for Work and Family: A Study of Women in Residential Real Estate Sales Work*. Lanham, MA: Lexington Books.

Whyte, William H. 1956. *The Organization Man*. New York: Simon & Schuster.

Wiesenfeld, Batia, Joel Brockner, and Christopher L. Martin. 1999. "A Self-Affirmation Analysis of Survivors' Reactions to Unfair Organizational Downsizing." *Journal of Experimental Social Psychology* 35(5):441–60.

Williams, Christine. 1991. *Gender Differences at Work: Women and Men in Non-traditional Occupations.* Los Angeles: University of California Press.

Williams, Christine, Patti Giuffre, and Kirsten Dellinger. 1999. "Sexuality in the Workplace: Organizational Control, Sexual Harassment, and the Pursuit of Pleasure." *Annual Review of Sociology* 25:73–93.

Williams, Damian. 2009. "Grounding the Regime of Precarious Employment: Homeless Day Laborers' Negatiation of the Job Queue." *Work and Occupations* 36:209–46.

Williams, Joan. 2000. *Unbending Gender: Why Family and Work Conflict and What to Do About It.* New York: Oxford University Press.

Wilson, George. 1997. "Pathways to Power: Racial Differences in the Determinants of Job Authority." *Social Problems* 44:38–54.

Wilson, Kenneth, and Alejandro Portes. 1980. "Immigrant Enclaves: An Analysis of the Labor Market Experiences of Cubans in Miami." *American Journal of Sociology* 86:295–319.

Wilson, William J. 1987. *The Truly Disadvantaged: The Inner City, the Underclass, and Public Policy.* Chicago: University of Chicago Press.

———. 1997. *When Work Disappears: The World of the New Urban Poor.* New York: Alfred A. Knopf.

Winslow-Bowe, Sarah. 2006. "The Persistence of Wives Income Advantage." *Journal of Marriage and Family* 68:824–42.

Winstanley, Diana, Joanna Clark, and Helena Leeson. 2002. "Approaches to Child Labour in the Supply Chain." *Business Ethics: A European Review* 11:210–23.

Wolkinson, Benjamin, and Russell Ormiston. 2006. "The Arbitration of Work-Family Conflicts." In *The Work and Family Handbook: Multi-disciplinary Perspectives, Methods, and Approaches,* edited by Marcie Pitt-Catsouphes, Ellen Ernst Kossek, and Stephen Sweet, 685–704. Mahwah, NJ: Lawrence Erlbaum.

Wood, Ellen Meiksins. 2003. *Empire of Capital.* London: Verso Books.

Woods, James, and Jay Lucas. 1993. *The Corporate Closet: The Professional Lives of Gay Men in America.* New York: Free Press.

Woods, Ngaire. 2006. *The Globalizers: The IMF, the World Bank, and Their Borrowers.* Ithaca, NY: Cornell University Press.

Wright, Eric Olin. 1985. *Classes.* London: Verso Books.

Wrigley, Julia. 1999. "Is Racial Oppression Intrinsic to Domestic Work? The Experiences of Children's Caregivers in Contemporary America." In *The Cultural Territories of Race: Black and White Boundaries,* edited by Michèle Lamont, 97–123. Chicago: : University of Chicago Press.

Wrong, Dennis. 1961. "The Oversocialized Concept of Man in American Sociology." *American Sociological Review* 26(2):183–96.

Yakura, Elaine. 2001. "Billables: The Valorization of Time in Consulting." *American Behavioral Scientist* 44:1076–95.

York, Richard, Eugene Rosa, and Thomas Dietz. 2003. "Footprints on the Earth: The Environmental Consequences of Modernity." *American Sociological Review* 68:279–300.

Young, Alford. 1999. "Navigating Race: Getting Ahead in the Lives of 'Rags to Riches' Young Black Men." In *The Cultural Territories of Race: Black and White Boundaries*, edited by Michèle Lamont, 30–62. Chicago: University of Chicago Press.

———. 2003. *The Minds of Marginalized Black Men: Making Sense of Mobility, Opportunity and Future Life Chances*. Princeton, NJ: Princeton University Press.

Zaretsky, Eli. 1986. *Capitalism, the Family, and Personal Life*. New York: Harper & Row.

Zelizer, Viviana. 1994. *Pricing the Priceless Child: The Changing Social Value of Children*. Princeton, NJ: Princeton University Press.

Zeng, Zhen, and Yu Xie. 2004. "Asian-Americans' Earnings Disadvantage Reexamined: The Role and Place of Education." *American Journal of Sociology* 109:1075–108.

Zimmer, Lynn. 1986. *Women Guarding Men*. Chicago: University of Chicago.

Zuboff, Shoshana. 1988. *In the Age of the Smart Machine*. New York: HarperCollins.

Zweigenhaft, Richard, and William Domhoff. 2003. *Blacks in the White Elite: Will the Progress Continue?* Lanham, MD: Rowman & Littlefield.

Index

activist groups and, 203–204
organized labor and, 208
time/work issues and, 122–123
See also Wages
Liz Claiborne, 66

Management:
mass production and, 26, 27
new cultures of control and, 35–36
Manpower Development and Training
    Act, 228
Manufacturing, 47
class inequalities and, 51
decline of, 63–64
India and, 70
job security and, 76
mass layoffs and, 83–84
organized labor and, 205
time/work issues and, 103–104, 117
unions and, 41 .*See also* Unions
in the U.S., 24–26, 25
Maquiladoras, 110
Marriages:
child care and, 153–154
job security and, 88–89
reshaping the new economy
    and, 197
time/work issues and, 119
working women and, 129–131 .
    See also Women
See also Families
Marx, Karl, 7–8, 13, 49, 200
Massachusetts Institute of Technology
    (MIT), 209
Mass layoffs, 83–84
Mass production, 26–30
McDonalds restaurants, 24, 29, 31–33
Medicaid, 217
Medicare, 217, 228
Mexican immigrants, 166, 190
immigration policy and, 189–190
Microloans, 65
Middle class:
affirmative action and, 187
African Americans and, 163
blue-collar jobs and, 12, 178
Chinese workers and, 69
developing countries and, 64, 66
low wages and, 55

race/ethnicity and, 161
racialized jobs and, 184–185
reshaping the new economy
    and, 197
time/work issues and, 110
See also Class; Immigration
Milkman, Ruth, 4
Minimum wage:
family care and, 158
FLSA and, 214
introduction of, 80
laboring in the new economy and, 5
race/ethnicity and, 172
role of government in the new
    economy and, 213
time/work issues and, 112, 122
See also Wages
Mining, 27–28
Minorities, 161–162, 222
affirmative action and, 185–187
careers/agency and, 20
cultural capital and, 177–178
geographic distribution of, 178–181
inequalities and, 196
Native Americans, 178–179
prejudice/discrimination and,
    181–184
reshaping the new economy and, 199
sexual, 146
social capital and, 175–176
social supports/job protections for, 78
See also African Americans; Asians;
    Ethnic minorities; Latinos;
    Women
Modern Times, 11
Multinational corporations:
globalization and, 46–47 .*See also*
    Globalization
human rights violations and, 211
Mutual assistance, 77

National Labor Relations Board, 48n 4
National Labor Relation (Wagner)
    Act, 227
Native Americans, 178–179
New economy, 2, 47–48, 222–223
activist groups and, 202–205
careers/agency and, 19–21
changes of incomes and, 51–53

SAGE research**methods**
The Essential Online Tool for Researchers

The essential tool for researchers . . .

. . . from the world's leading research methods publisher

Discover SRMO Lists—methods readings suggested by other SRMO users

"I have never really seen anything like this product before, and I think it is really valuable."

John Creswell, University of Nebraska–Lincoln

Find exactly what you are looking for, from basic explanations to advanced discussion

Explore the Methods Map to discover links between methods

Watch video interviews with leading methodologists

Search on a newly designed taxonomy with more than 1,400 qualitative, quantitative, and mixed methods terms

Uncover more than 100,000 pages of book, journal, and reference content to support your learning

find out more at
srmo.sagepub.com